Blockchain for 6G-Enabled Network-Based Applications

This book provides a comprehensive overview of blockchain for 6G-enabled network-based applications. Following the key services of blockchain technology, this book will be instrumental to ideate and understand the necessities, challenges, and various case studies of different 6G-based applications. The emphasis is on understanding the contributions of blockchain technology in 6G-enabled applications, and its aim is to give insights into evolution, research directions, challenges, and the ways to empower 6G applications through blockchain.

- The book consistently emphasizes the missing connection between blockchain and 6G-enabled network applications. The entire ecosystem between these two futuristic technologies is explained in a comprehensive manner.
- The book constitutes a one-stop guide to students, researchers, and industry professionals.
- The book progresses from a general introduction toward more technical aspects while remaining easy to understand throughout.
- Comprehensive elaboration of material is supplemented with examples and diagrams, followed by easily understandable approaches with regard to technical information given thereon.

Blockchain and its applications in 6G-enabled applications can drive many powerful solutions to real-world technical, scientific, and social problems. This book presents the most recent and exciting advances in blockchain for 6G-enabled network applications. Overall, this book is a complete outlet and is designed exclusively for professionals, scientists, technologists, developers, designers, and researchers in network technologies around blockchain integration with IoT, blockchain technology, information technology, and 6G-enabled industrial applications. Secondary readers include professionals involved in policy making and administration, security of public data and law, network policy developers, blockchain technology experts, regulators, and decision makers in government administrations.

T0321383

Blockchain for 6G-Enabled Network-Based Applications

A Vision, Architectural Elements, and Future Directions

Edited by
Vinay Rishiwal
Sudeep Tanwar
Rashmi Chaudhry

CRC Press

Taylor & Francis Group

Boca Raton London

CRC Press is an imprint of the
Taylor & Francis Group, an **informa** business

First Edition published 2023
by CRC Press
6000 Broken Sound Parkway NW, Suite 300, Boca Raton, FL 33487–2742

and by CRC Press
4 Park Square, Milton Park, Abingdon, Oxon, OX14 4RN

CRC Press is an imprint of Taylor & Francis Group, LLC

ISBN: 978-1-032-20610-3 (hbk)
ISBN: 978-1-032-20611-0 (pbk)
ISBN: 978-1-003-26439-2 (ebk)

DOI: 10.1201/9781003264392

Typeset in Sabon
by Apex CoVantage, LLC

Contents

Preface

The emergence of smart wireless and mobile communication technologies has stirred the demand for future large-scale networks. Sixth-generation (6G) networks seem to be the possible solution to meet such demands of developing networks and scenarios like IoT, smart cities, unmanned vehicles, satellite communication, etc. 6G wireless communication offers improved data rate, reliability, and bandwidth utilization. Until now, four core services of 6G have been identified: enhanced mobile broadband (eMBB) and massive machine-type communications (mMTC), ultra-reliable low latency communications (URLLC), enhanced URLLC, and trade-offs among them. It aims to satisfy the Internet of Everything (IoE) requirements in step with the growth of distributed artificial intelligence while maintaining ultralow latency and device synchronicity. Such evolutionary improvements introduce better connectivity and novel applications focusing on privacy and security. To cater to the explosive growth of data traffic and seamless connection requirements, virtualization and cloud computing are used, which provide network slicing and vertical services. However, the management of such data in a centralized manner induces privacy and security issues. Recently, blockchain technology has gained momentum to solve such data privacy, reliability, security, and vulnerability issues, and it is being embraced by research communities and the industry. It provides decentralization, single point of failure elimination, anonymous transparency, immutability, tamper-proofed ledger's data, and nonrepudiated transactions. Blockchain-assisted distributed ledger technology (DLT) helps in building trust between application-specific networks and IoE applications.

In order to get the expected benefits of these two futuristic technologies. The primary requisite is to have enough imaginings and to get ready and lay a sound technological base for the possible arrival of future networks. To summarize, it is the right time to begin exploring next-generation 6G mobile services/networks. Following the key services of blockchain technology in 6G network–based applications, this book focuses on the key facets of these groundbreaking technologies. It helps to understand the ideas, necessities, challenges, and, largely, the overall framework of blockchain and 6G networks. This book focuses on the exploration of the core benefits of blockchain in 6G network–based applications and provides an overview of the integration of blockchain with 6G-enabled network-based applications.

The book is organized into three sections. The first section is focused on the background and preliminaries and includes three chapters. The second part illustrates AI-assisted secure 6G-enabled solutions with three well structured chapters. Finally, the last section, with four chapters, focuses on the next-generation 6G-enabled solutions for smart applications.

SECTION 1: BACKGROUND AND PRELIMINARIES

The first chapter, "Evolution and Innovation of Blockchain Paradigms for the 6G Ecosystem," presents an introduction to blockchain and the 6G ecosystem. The chapter discusses the blockchain paradigms with 6G technology in various realms, with innovative decentralized architecture in multicross industries and enhances the security access in new integrated network architecture like IoT, device-to-device (D2D), machine-to-machine (M2M), smart city, artificial intelligence, and machine learning (ML).

Chapter 2, "Empowering 6G Networks through Blockchain Architecture," highlights the background and research challenges of 6G networks through blockchain architecture. Moreover, this chapter explores the future challenges of network improvement and scaling through the inclusion of blockchain technology and NFV concerning the following 6G service classes: secure, ultra-reliable low-latency communication (SURLLC), three-dimensional integrated communications (3D InteCom), eMBB Plus, unconventional data communications (UCDC), big communications (BigCom), massive Machine Type Communication mMTC.

The next chapter, "Blockchain Technology and 6G: Trends, Challenges, and Research Directions," discuss the role of blockchain technology in 6G, applications of blockchain technology in real time, and the trends, challenges, and research directions for blockchain technology. The combination of 6G wireless networks and blockchain addresses the multiple network issues related to real-time applications. The proposed work is explained by a case study of the logistics supply chain. This chapter also highlights the benefits of a 6G network with blockchain technologies.

SECTION 2: AI-ASSISTED SECURE 6G-ENABLED SOLUTIONS

In Chapter 4, "Blockchain and Artificial Intelligence–Based Radio Access Network Slicing for 6G Networks," radio access network (RAN) slicing is highlighted as a key enabler for 6G wireless communications that can flexibly manage base stations, network functions, network resources, bandwidth, and network configurations to accommodate the diverse conditions and requirements imposed by the emerging use-cases. First, key 6G driving use-cases, including 5D applications, smart cities and industries, and nonterrestrial communications will be elucidated. Then an overview of network slicing for wireless communications will be presented, emphasizing RAN slicing. Subsequently, a detailed discussion on RAN slicing for 6G networks, including its key aspects, requirements, parameters, and scenarios, will be provided. Next, several artificial intelligence (AI) technologies that can assist the complex RAN slicing process with numerous parameters and target requirements will be highlighted. Then blockchain-based network slicing will be explored for 6G networks, followed by a case study on the integration of AI and blockchain for 6G RAN slicing.

The next chapter, "Key Generation: Encryption and Decryption Using Triangular Cubical Approach for Blockchain in the 6G Network and Cloud Environment," proposes a cubical key process for key generation using a two-dimensional platform and a key matrix, which are used to generate from combinations and permutations, as well as constructing encryption and decryption keys to apply to encryption keys, plaintext, and ciphertext. This is used in the blockchain of 64 integer messages as different nonbinary ciphers to increase the security of encryption and decryption in the 6G network and cloud computing environment. This chapter discusses the process with an example, tests the proposed method, and checks the entropy, including brute attack aspects.

Chapter 6, "Blockchain-Based Efficient Resource Allocation and Minimum Energy Utilization in Cloud Computing Using SMO," presents a blockchain-based resource management framework and spider monkey optimization (SMO) for resource allocation, enhancing scheduling ability and overall resource allocation in a cloud computing environment. To increase energy efficiency, the brownout technique is adopted. The proposed approach is simulated using the CloudSim tool kit. This framework achieves cloud infrastructure resource allocation and reduced power consumption. The experimental outcomes provide improved results when compared to other prevailing techniques.

SECTION 3: NEXT-GENERATION 6G-ENABLED SOLUTIONS FOR SMART APPLICATIONS

The first chapter in the section, "Role of Blockchain Technology in Intelligent Transportation Systems," presents how blockchain technology helps in improving reliable vehicular communication. This chapter also discusses the overview of intelligent transportation systems, VANET, and blockchain technology. It also depicts the system model with three main facets: the onboard unit in a conveyance, roadside unit of a network, and trusted authority of the network for vehicular communication. Security attacks based on confidentiality between nodes, availability of resources, authenticity of conveyances, integrity of messages, and nonrepudiation of nodes are also illustrated. The last section explains the bulwark vehicular memorandum using blockchain technology.

Chapter 8, "Blockchain and 6G Communication with Evolving Applications: Forecasts and Challenges," briefly deliberates upcoming challenges in 6G communication for canvassers. Various 6G application areas are also presented in this survey. The key focus is associated with 6G architecture, which includes protocols stacks, coverage, and artificial intelligence. The goal of this chapter to give informative direction for consequent 6G communication research.

The next chapter, "Emergence of Blockchain Applications with the 6G-Enabled IoT-Based Smart City," offers an expository view of the implementation of blockchain technology in connection with IoT, 5G, and artificial intelligence. Also, the architecture implementation of blockchain in assimilation with IoT and the role of artificial intelligence in automating the processes and establishing a user-friendly standard operating procedure for smart cities have been described. A different generation of networks that support IoT and act as its backbone is explained lastly.

Finally, Chapter 10, "Blockchain: Insights Related to Application and Challenges in the Energy Sector," aims to identify and utilize the attributes of blockchain technology in the energy domain beyond its basic use. The chapter will first define the use of blockchain in application areas of the energy domain. Later on, the chapter highlights various challenges in the energy sector due to the adaptation of blockchain. The role of blockchain in paving the path for turning these challenges into opportunities is also a major focus.

Dr. Vinay Rishiwal
Bareilly, India
Dr. Sudeep Tanwar
Ahmedabad, India
Dr. Rashmi Chaudhary
Raipur, India

About the Editors

Dr **Vinay Rishiwal** works as Professor in the Department of Computer Science and Information Technology, Faculty of Engineering and Technology, MJP Rohilkhand University, Bareilly, Uttar Pradesh, India. He obtained a BTech degree in Computer Science and Engineering in 2000 from MJP Rohilkhand University (SRMSCET), India, and received his PhD in Computer Science and Engineering from Gautam Buddha Technical University, Lucknow, India, in 2011. He has 20 years of experience in academics and is a senior member of IEEE and has worked as Convener Student Activities Committee, IEEE Uttar Pradesh Section. He has published more than 90 research papers in various journals and conferences of international repute. He is a general/conference chair of four international conferences: ICACCA, IoT-SIU, MARC 2020, and ICAREMIT. He has received many awards for best paper and best orator in various conferences. Dr. Rishiwal has visited many countries for academic purposes and has worked on many projects of MHRD and UGC. His current research interest includes wireless sensor networks, IoT, cloud computing, social networks and blockchain technology. He was invited as guest editor/editorial board member of many international journals and has been invited as keynote speaker in many international conferences held in India and abroad.

Dr **Sudeep Tanwar** works as a full Professor in the Computer Science and Engineering Department at Institute of Technology, Nirma University, India. Dr Tanwar is a visiting professor in Jan Wyzykowski University in Polkowice, Poland, and in the University of Pitesti in Pitesti, Romania. Dr Tanwar's research interest includes blockchain technology, wireless sensor networks, fog computing, smart grid, and IoT. He has authored two books and edited 13 books and has published more than 270 technical papers, in top journals and at top conferences, such as IEEE TNSE, TVT, TII, WCM, Networks, ICC, GLOBECOM, and INFOCOM. Dr Tanwar initiated the research field of blockchain technology adoption in various verticals in 2017. His h-index is 49. Dr Tanwar actively serves his research communities in various roles. He is currently serving on the editorial boards of *Computer Communications, International Journal of Communication System, Cyber Security and Applications, Frontiers in Blockchain,* and *Security and Privacy.* He has been awarded best research paper awards from IEEE GLOBECOM 2018, IEEE ICC 2019, and Springer ICRIC-2019. He has served many international conferences as a member of organizing committee, such as publication chair for FTNCT-2020, ICCIC 2020, WiMob 2019; a member of advisory board for ICACCT-2021, ICACI 2020; workshop cochair for CIS 2021; and general chair for IC4S 2019, 2020, ICCSDF 2020. Dr Tanwar is a final voting member for the IEEE ComSoc Tactile Internet Committee in 2020. He is a member of IEEE, CSI, IAENG, ISTE, CSTA, and the Technical Committee on Tactile Internet of IEEE Communication Society. He leads the ST research lab where group members are working on the latest cutting-edge technologies

Dr Rashmi Chaudhry is working as Assistant Professor in the Computer Science and Engineering Department at Dr Shyama Prasad Mukherjee International Institute of Information Technology Naya Raipur (IIITNR), Sector 24, Naya Raipur, Chhattisgarh. Prior to joining IIITNR, she was a faculty member in TIET, Patiala, Punjab, India. She received her PhD from ABV-IIITM, Gwalior, and her MTech from IIT (Indian School of Mines), Dhanbad, India. She has been a visiting research scholar at Anglia Ruskin University, Chelmsford, UK, in 2015 under the UKIERI project between India and the UK. She has published research papers in leading journals and conferences of IEEE, Elsevier, and Springer. Her current interest includes wireless sensor networks, IoT, fog computing, machine learning and intelligent transportation systems.

Contributors

CHAPTER 1

R. Durga
Department of Computer Science and Engineering
SRM Institute of Science and Technology
Kattangulathur, India

E. Poovammal
Department of Computer Science and Engineering
SRM Institute of Science and Technology
Kattangulathur, India

G. Kothai
Department of Computer Science and Engineering
SRM Institute of Science and Technology
Kattangulathur, India

CHAPTER 2

Kaustubh Ranjan Singh
Department of Electronics and Communication Engineering
Delhi Technological University
Delhi, India

Parul Garg
Department of Electronics and Communication Engineering
Netaji Subhas University of Technology
Delhi, India

CHAPTER 3

Shyam Mohan
Faculty in the Department of CSE
Sri Chandrasekarendra Saraswathi Viswa Maha Vidyalaya (SCSVMV)
Enathur, India

S. Ramamoorthy
Faculty in the Department of CSE
SRM Institute of Science and Technology
Kattankulathur, India

Vedantham Hanumath Sreeman
Final Year CSE
Sri Chandrasekarendra Saraswathi Viswa Maha Vidyalaya (SCSVMV)
Enathur, India

Venkata Chakradhar Vanam
Final Year CSE
Sri Chandrasekarendra Saraswathi
    Viswa Maha Vidyalaya (SCSVMV)
Enathur, India

Kota Harsha Surya Abhishek
Final Year CSE
Sri Chandrasekarendra Saraswathi
    Viswa Maha Vidyalaya (SCSVMV)
Enathur, India

CHAPTER 4
Ying Loong Lee
Department of Electrical and Elec-
    tronic Engineering
Lee Kong Chian Faculty of Engineer-
    ing and Science
Universiti Tunku Abdul Rahman
Selangor, Malaysia

Allyson Gek Hong Sim
Secure Mobile Networking
    (SEEMOO) Lab, Department
    of Computer Science
Technical University
    of Darmstadt
Darmstadt, Germany

Li-Chun Wang
Department of Electrical and
    Computer Engineering
National Chiao Tung University
Hsinchu, Taiwan

Teong Chee Chuah
Faculty of Engineering
Multimedia University
Selangor, Malaysia

CHAPTER 5
Surendra Kumar
Department of Computer Science
Babasaheb Bhimrao Ambedkar Uni-
    versity (A Central University)
Lucknow, India

Narander Kumar
Department of Computer Science
Babasaheb Bhimrao
    Ambedkar University
    (A Central University)
Lucknow, India

CHAPTER 6
Narander Kumar
Department of Computer Science
BBA University (A Central
    University)
Lucknow, India

Jitendra Kumar Samriya
Department of Computer Science
BBA University (A Central
    University)
Lucknow, India

CHAPTER 7
G. Kothai
Department of Computer Science
    and Engineering
SRM Institute of Science and
    Technology
Kattankulathur, India

E. Poovammal
Department of Computer Science
    and Engineering
SRM Institute of Science and
    Technology
Kattankulathur, India

R. Durga
Department of Computer Science
    and Engineering
SRM Institute of Science and
    Technology
Kattankulathur, India

CHAPTER 8
Preeti Yadav
Department of CS & IT
MJP Rohilkhand University
Bareilly, India

Mano Yadav
Department of Computer Science
Bareilly College
Bareilly, India

Omkar Singh
Department of CS & IT
MJP Rohilkhand University
Bareilly, India

Vinay Rishiwal
Department of CS & IT
MJP Rohilkhand University
Bareilly, India

CHAPTER 9
Meenu Gupta
Department of Computer Science
and Engineering
Chandigarh University
Punjab, India

Chetanya Ved
Department of Information
Technology
Bharati Vidyapeeth's College of
Engineering
Delhi, India

Meet Kumari
Department of Electronics and
Communication Engineering
Chandigarh University
Punjab, India

CHAPTER 10
Iram Naim
Department of Computer Science &
Information Technology
MJP Rohilkhand University
Bareilly, India

Ankur Gangwar
Department of Computer Science &
Information Technology
MJP Rohilkhand University
Bareilly, India

Ashraf Rahman Idrisi
Department of Maintenances and
Services
Bharat Heavy Electrical Limited
New Delhi, India

Pankaj
Department of Computer Science &
Information Technology
MJP Rohilkhand University
Bareilly, India

Chapter 1

Evolution and Innovation of Blockchain Paradigms for the 6G Ecosystem

R. Durga,[1] E. Poovammal,[2] and G. Kothai[3]

CONTENTS

[1,2,3] Department of Computer Science and Engineering, SRM Institute of Science and Technology, Kattangulathur, India
Corresponding author: dr8529@srmist.edu.in

DOI: 10.1201/9781003264392-1

1.1 INTRODUCTION

The wireless telecommunication network mutates nearly every ten years, giving meteoric innovations since 1980. In 1981, the first generation of wireless cellular technology was analog standard radio signals. Then 2G digital networks replaced analog radio signals. The second-generation cellular networks are time division multiple access (TDMA)-GSM standards consisting of three primary benefits: conversations between users are digitally encrypted, methodical usage of radio frequency, and new data services mode on short messaging service (SMS) text.

In 2001, the third generation of wireless mobile telecommunication technology was upgraded from 2G for faster data transfer using International Mobile Telecommunication Standards. In 2011, the fourth generation of broadband cellular technology, using the International Mobile Telecommunication Advanced Specification, relied on Internet protocol-based communication. The fourth generation was usually described as long-term evolution (LTE), which offers highly defined mobile television, video conferencing, Internet protocol telephony, and 3D television [1]. In 2019, the fifth-generation technology standard for cellular networks was deployed worldwide. The 5G networks had greater bandwidth than 4G, which offered a significant advantage in providing higher download speeds [2]. Due to the growth in bandwidth, there was an increase in applications in machine-to-machine networks and the Internet of Things.

1.1.1 Generations before 6G

The sixth-generation network is the revolutionary innovation of this decade. At all previous levels of generation, the network had been vulnerable to security threats. The first generation wireless networks were elementary and vulnerable to illegal cloning and many security exploitations, such as masquerade attacks. The second generation was susceptible to message spamming, such the injection of false information into the networks. The third generation of wireless networks consisted of many Internet-based vulnerabilities in IP-based protocols. The fourth generation deployed many new multimedia, and hence attacks increased on smart devices [1]. The fifth generation, with many interconnected critical services, was expanded more than 4G, increasing the threat of actors in network applications, such as smart cities and smart transport communication [1]. See Tables 1.1 and 1.2.

The areas of study in the standardized development in terahertz wireless communication (THZ) [3] and visible light communication (VLC) are not included in the 5G cellular network, but innovative research is in progress to track non-line-of-sight environments. In 5G, cellular networks sensing using RF signals are still in research, not integrated for operations. But 6G networks

Table 1.1 Comparison of Generations

Features	1G	2G	3G	4G	5G	6G
Start	1970– 1984	1980– 1999	1990– 2002	2000–010	2010– 2015	2019
Technology	AMPS, NMT, TACS	GSM	WCDMA	LTE, WiMAX	MIMO, millimeter waves	GPS
Frequency	30 kHz	1.8 GHz	1.6–2 GHz	2–8 GHz	3–30 GHz	95 GHz–3 THz
Access system	FDMA	TDMA/ CDMA	CDMA	CDMA	OFDM/ BDMA	COMPASS
Core network	PSTN	PSTN	Packet networks	Internet	Internet	Internet
Speed	2.4 Kbps	64 Kbps	2 mbps	100 mbps	300 mbps	1 Tbps
Handoff	Horizontal	Horizontal	Horizontal	Horizontal/ Vertical	Horizontal / Vertical	Horizontal/ Vertical
Service	Mobile telephony	Digital voice, short messaging	Integrated high audio and video	Dynamic information access	Dynamic information access on variable devices	Dynamic information access on variable devices with AI capabilities
Switching	Circuit	Circuit for access N/W and air interface	Packet except air interface	All packets	All packets	All packets
Satellite integration	No	No	No	No	No	Yes
AI	No	No	No	No	Partially	Fully
E2E latency	No	No	No	100 ms	10 ms	1 ms
Mobility support	No	No	No	Up to 350 km/h	Up to 500 km/h	Up to 1000 km/h

Table 1.2 Distinctive Differences between 5G and 6G

S.NO	Standards	Technological limitations in 5G	6G challenges	Supported use-cases
1	Terahertz	High bandwidth with Small antenna	High propagation loss	Industry 4.0
2	VLC	Unlicensed spectrum with low interference	RF uplink needed	E-health
3	Sensing and localization	Context-based control	Efficient multiplexing	Unmanned mobility
4	User-based network architecture	Distributed intelligence	Real-time process	Pervasive connectivity, e-health

rely on beam foaming patterns to facilitate operations on e-health and intelligent vehicular networks as innovative services. The 5G networks, which provide bidimensional space, offer connectivity to the device and ground. 6G provides three-dimensional coverage through terrestrial and nonterrestrial platforms by drones and satellites. Thereby, the functional standards of 6G meet the requirements of ultrahigh data rates and high-volume traffic while envisioning essential growth in blockchain technology.

1.1.2 Motivation and Scope of Blockchain in 6G

Blockchain technology is generally a tamperproof, decentralized database ordered in a hash tree structure. The features of blockchain technology are atomicity, durability, authenticity, suitability, integrity, privacy, and security. The benefits of blockchain technology [4] to 6G are (1) decentralization that enables zero-touch service management and improves the service strength in the nodes of the network; (2) immutability, which creates a trusted execution environment for the networks made of unbounded interconnected devices; (3) elimination of single-point failure by adopting standard cryptographic algorithm/advanced cybersecurity techniques; (4) smart contracts that enable data security, sharing, and service-level agreements; and (5) lower processing fees and minor processing delay for large-scale data exchange compared to 5G.

1.1.3 Contribution of Blockchain Using 6G in Various Realms

This chapter foresees that the future pathway for blockchain framework use in 6G technology is in enhancing security features in various realms such as IoT, device-to-device (D2D), machine-to-machine (M2M), smart contract, artificial intelligence and machine learning (ML). These highlighted enhanced security features of blockchain intelligent resource management employ smart contracts for spectrum sharing and provide self-organizing networks.

This chapter is organized as follows. Section 1.2 explains how 6G enhances secured access by inheriting the blockchain framework. Section 1.3 debates how 6G was restructured from 5G in terms of data rates and latency used in multicross industries. Section 1.4 discusses how 6G utilizes decentralized architectures on applications like automated robotic systems, augmented reality, and 3D printing. Section 1.5 describes the standards of blockchain adopted for 6G technology. Section1.6 summarizes the blockchain framework empowered by 6G for the new integrated network architecture for intelligent-based, device-to-device communication. Section 7 concludes by unwrapping the challenges of the 6G-based blockchain ecosystem.

1.2 ENHANCED DATA SECURITY IN 6G NETWORKS

The 6G network is a revolutionary innovation of this decade. The applicable standards of 6G meet the requirements of ultrahigh data rates and high-volume traffic. The 6G system provides a new structure from the existing generation. However, excellent AI services will soon design the architecture and operations of 6G. In March 2019, 6G wireless cellular communication was framed at the symposium with the vision statement "The Wireless Intelligence" [5]. The vision statement focused on four critical features like multiband fast transmission, energy-efficient communication, artificial intelligence, and high security and privacy.

The 6G technology has enhanced data security by inheriting the potential uses of blockchain technology: decentralization behavior in networks, the distributed ledger concept, spectrum sharing. Different security requirements need to be met by 6G technology:

Confidentiality: The immutable, tamperproof cryptographic techniques of blockchain can enhance data confidentiality in 6G networks [6].

Integrity: The abounding data produced by 6G networks requires that it be attainable by the authenticated users in transit. The hash function in blockchain technology helps 6G to prevent eavesdropping and modifying data while it is in transit and thus change the functionality of the system.

Authentication and access: Storing data in a decentralized, distributed manner gives future consumers in 6G networks sophisticated access mechanisms and precludes the possibility of single point of failure.

Audit: In 6G networks, auditing challenges security since it involves many tenants. The decentralized, distributed nature of blockchain is immutable, and the transparent log helps in auditing the events happening in 6G networks.

When it comes to 6G systems, blockchain technology is a game changer since it provides a level of security and anonymity that is undisputed by any other technology.

1.2.1 Threats in 5G Networks

In 5G networks, the security and privacy-preserving issues crop up at the architecture tiers: the core network tier, the backhaul network tier, and the access network.

In the core network tier, the network function virtualization (NFV) and software-defined networking (SDN) techniques are the kinds of functions that produce high vulnerabilities like DoS and resource attacks [7].

In the backhaul network tier, the communication done across the base station to the core networks by wireless medium or wired channels leads to the high risk of security issues.

Apart from these tier threats, significant threats occur at the components assigned at the edge of 5G networks, called multiedge computing threats.

Within the 5G infrastructure, there are threats accompanying virtualization functions and networks called virtualization threats. The physical threats also occurring in the infrastructure lead to damage, equipment loss, or failure of equipment and malfunctions in 5G Networks.

1.2.2 Requirement Syntax of Blockchain in 6G

Blockchain technology, with its elevated security features, unleashes the capacity of 6G systems in terms of privacy, authentication, access control, integrity, accountability, availability, and scalability. Preferment security aspects are as follows:

Privacy: Privacy is the consequential deliberation of security. In diverse 6G networks, data privacy becomes an aggravated security issue. A content-centric privacy-preserving system in blockchain technology was proposed by Fan et al. [8].

Authentication and access control: In centralized servers, the issue is limitation of scalability. The authentication access mechanism in the blocking technology was introduced by Yang et al. [9] for fiber networks.

Integrity: The 6G networks, with their ultrahigh data rates, can generate questions regarding data integrity. To overcome the data integrity issue, the blockchain framework was proposed by Ortega et al. [10] to exchange information over networks.

Availability: The availability of 6G services will play a predominant role in future wireless communication. Thus many security attacks, like DDoS, is possible due to the high connectivity in 6G networks. The blockchain-based DDoS prevention mechanism was proposed by Rodrigues et al. [11] for the protection of data.

Accountability: The accountability of 6G technology–based networks is a crucial requirement. The distributed ledger technique in the blockchain framework stores all the details in an ordered manner, satisfying the essential requirement [4].

Scalability: In 5G networks, scalability is the primary concern due to the centralized server structure. The smart contract in blockchain and decentralized distributed feature precludes the limitation on scalability.

1.3 DECENTRALIZED INTELLIGENT RESOURCE MANAGEMENT

To restructure the shape of 5G in terms of reliability, data rates, and latency, several lines of research activities have been initiated that show how 6G meets super-requirements:

1. 6G provides at least 1 Tb/s and low latency of 1 ms [12].
2. 6G increases the efficiency of the network spectrum and operates in the frequency range of GHz to THz.
3. It can connect a high number of things, enabling the Internet of Everything.

In mobile networks, resource management and spectrum assignment are the challenging paradigms. The 6G technology follows revolutionary wireless communication that overcomes the threats and achieves many new applications on device–to-device, mobile edge computing, the Internet of Everything [13].

The primary objective of 6G technology is to integrate the three satellite types, a.: (1) telecommunication multimedia networks, (2) navigation, and (3) weather information services. The 6G technology, with its combination of artificial intelligence and nanocore, uses integrated applications for Earth imaging satellites to monitor the resources for information about weather. The most attractive feature of 6G technology is managing a heavy volume of data and the soaring data rate connectivity per device. By 2030 [14], 6G technology development will meet the requirements of consumers with high capacity, ultrahigh efficiency, very low latency, high scalability, and high throughput is depicted in Figure 1.1.

1.3.1 Application-Based Use-Cases in 6G Networks

The blockchain coupled with 6G will verify all the peers and record all the events across device-to-device communication with its associated timeframe [15].

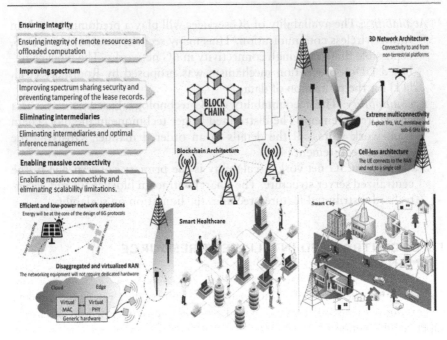

Figure 1.1 Decentralized 6G architecture for future applications.

Blockchain technology deals with the new challenges on dynamic spectrum management, and it acts as a catalyst to improve the management of convolutional spectrum and spectrum auction [16, 17]. Based on the block framework, the different challenges are highly lightened in IoT, D2D, network slicing and communication, Spectrum Computing, Energy trading

1.3.1.1 Spectrum Management

To meet the flourishing demand of massive data rate, network capacity needs improvement for further applications to be developed in 6G. Hence there is a high demand in the spectrum. The active policy of spectrum management has been proposed recently to utilize the range efficiently and effectively. The purpose of the dynamic policy [18] is that unlicensed secondary users can access the licensed spectrum without obstructing the licensed users. This application started to rely on the blockchain as the promising database for various kinds of opportunities: sensing the spectrum, the outcomes of data mining, results of spectrum auction, mapping of spectrum leashing [19]. The data about the spectrum is recorded in the blocks in an immutable manner. As spectrum management anticipates recording the data, along with the blockchain, the primary user's database was protected in TV white space (TVWS). Prior users can control their data in the blocks interlinked as a chain structure. So that logged spectrum such as time, frequency, TVWS geolocation will be recorded. Access control can be achieved in the blockchain using smart contracts consisting of scripts based on terms and conditions. The threshold is maintained in the smart contract so that the users cannot access the band after a particular time [20]. Every user is retained with the access history, whereas in traditional carrier-sense multiple access (CSMA) schemes, user access is not defined correctly.

1.3.1.2 Pervasive Connectivity

The increase in traffic growth in wireless communication is expected to increase three times the 2016 level, with 10^7 devices per kilometer in a densely populated region. In contrast, 125 billion device-to-device communications are expected by 2030. The future requirement includes scalable, low-cost deployment with low impact on the environment for better coverage. The pervasive connectivity in 6G overcomes the challenges faced by 5G networks that work in the millimeter-range wave spectrum, rendering indoor connectivity less effective due to the inability to penetrate appropriately through concrete material.

1.3.1.3 Network Slicing

Vertical industry services had diverse requirements on network slicing for upcoming cellular architecture. VNF (virtual network function) is clubbed as a chain structure on massive machine-type communication with enhanced mobile broadband at ultra-reliable low latency to realize network slicing [21]. The robot-based surgery on orchest and management of network slicing with the required high security need to be tested in vehicle-to-everything (V2X) communication to accommodate the applications.

The most critical challenge of network slicing is that the available resources in the network be transparent with several players. The integration of 6G and blockchain technology functionalities laid the path for the trust to make network slicing applications. A resource brokerage is an option that the network could take as a lease-on-demand requirement [22]. Smart contracts obtain this kind of trading in network brokerage, and cryptographic techniques in blockchain eradicate the single point of failure possibility and ensure security. Once the data is stored inside the blocks per a contract of agreement, it can be accessed through smart contracts [23]. The blockchain technology in 6G gives gain points on resource brokerage like operational cost saving, low-cost slicing agreements, enhanced security on slice transactions, high efficiency on network slicing [20].

By anchorage, the disintermediate blockchain on network slicing will offer remote surgery and controlling the drones remotely [24]. The decentralized-based approach provides automatic reconciliation and the payment between the providers in different locations, cutting the cost for third-party interven-tion. The use-case of this application is discussed by the author [20, 25], which builds the blockchain-based model for automating payment and for billing through distributed ledger by managing the rules and protocols. The model incorporates all these operations in the service layer, processed through the smart contract.

Thus, using the blockchain technology for automated brokerage in slicing the network offers gains on many positive points, such as economizing on operational cost, accelerating slicing and negotiation, and thereby simplifying the agreement and enhancing security in a transaction during slicing.

1.4 DECENTRALIZED ARCHITECTURE FOR INDUSTRIAL APPLICATIONS 4.0

Industrial applications 4.0 will facilitate efficient operations via smart supply chains and smart factories. This digitalization era will ultimately lead to creat-ing innovative models for business and to offering new services. Blockchain technology makes it practicable by fusing the physical and cyber systems for Industry 4.0–based platforms like IoT, 3D printing, augmented reality, and business models [26]. These emerging business models are being structured in end-to-end design, interconnected using secured blocks in the blockchain framework.

The blockchain platform is a capability framework that adopts healthcare, government, education, transport and logistics applications. In healthcare applications, the blockchain framework can apprehend a patient's lifetime his-tory of medical reports since a permission network can assure privacy and exhibits confidentiality by agreement through a smart contract. The govern-ment is turning its priority toward the potential of blockchain technology for the betterment of services to citizens. It facilitates the automatic administra-tive processes by excluding fraud, thereby improving trust and confidentiality through immutable distributed ledgers.

The fundamental approach of blockchain features is to circumvent fraud certification, turning educational institutions toward the decentralized data-base platform. The responsible social environment that absorbs the block-chain technology platform provides many opportunities for logistics in terms of tracking, cost, and time saving in the supply chain sectors [27].

Industry 4.0 plays a vital role in core drivers, such as smart solution-based IoT, smart products-based robotics and automated systems, smart supply chain, smart factory with innovative applications on augmented reality, 3D printing, and cyber security. The prominent features of 6G and blockchain technology enable Industry 4.0 to provide a significantly higher grade of trans-parency, security, and efficiency in interdisciplinary sectors in the areas of data science, health innovations, smart city, and many social innovation products by using the unique features of the smart contract depicted in Figure 1.2 [28]. Some of the sectors are elaborated in the following subsections.

1.4.1 Automated Robotic Systems

Automated robotic systems are autonomous designs that can move all by themselves using sensors in a dynamic environment. The machine-to-machine must collaborate and understand the unknown environment by studying the inputs from that environment. With the 6G, the Industry 4.0 application revolution started with the cyber-physical system architecture. Since 6G technology had a high Gbps peak data rate with the speed of 1 Tbps with a low latency of 1 ms, it offers the complete autonomous package for the transport system. The real-time scenarios of automated flying drone systems [28] using the Internet could work on the improved capacity of 6G technology and of breakthrough network models' advancements through blockchain technology.

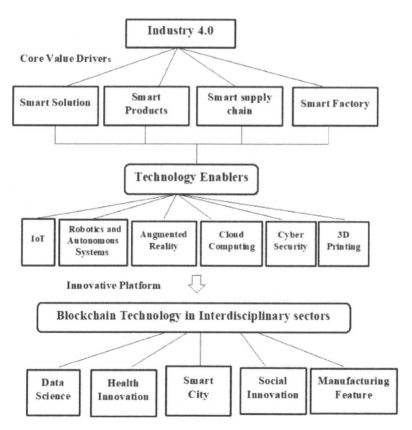

Figure 1.2 Prominent features of 6G and blockchain-enabled Industry 4.0 sectors [28].

1.4.2 Augmented Reality

A device that can display elements virtually in the physical, real-world environment is said to be capable of augmented reality. Augmented reality and virtual reality help improve industry-based perspectives like production increment and maintenance of components, specifically industry-based-designed equipment. The primary criteria for augmented and virtual reality devices are sensing real data through sensors and storing that data on the server for further tracing processing. So wearable devices that use augmented and virtual reality exchange GPS location data over the Internet, which may cause threats such as third parties accessing secured information. In this situation, blockchain technology can be used to store the data in the blockchain. 6G technology uses a high speed of exchanging data at 95 GHz to 3 THz, using GPS technology to locate the data quickly.

Augmented and virtual reality devices need considerable bandwidth for data transactions for asynchronous communication with the server. Due to high data transactions, the data may overload the system, and services may be delayed. The 6G network that combined the blockchain helps boost data availability, and the decentralized ledger helps to record real-time data sequentially in blocks. The decentralized design for augmented and virtual reality devices allows primary users and innovative developers to store, upload, download, and create their marketplace like a decentralized ecosystem [29, 30]. The financial applications of augmented and virtual reality devices can act as the wallet for peer-to-peer digital token transactions [31].

1.4.3 3D Printing Additive Manufacturing Unit

In the Industry 4.0 application, customization and flexibility are the two paradigms for innovative factory development. Apart from the traditional techniques for manufacturing, 3D printing permits some prototypes for manufacturing in the low-volume sectors at a meager cost with the aid of the blockchain. The decentralized just-in-time factory 4.0 was implemented globally through automated smart contracts by the correlation of 3D printers via blockchain technology [32]. Validation and authentication for testing an agent are necessary for manufacturing because many providers and suppliers are actors in the files of the manufacturing Industry 4.0. The unique design was a concern in additive manufacturing in Industrial Application 4.0. Since many actors accessing the files can copy the design template and replicate it for an infinite number of copies, a simple smart contract can act as a licensing agreement, can be made, so that the documents belong to the particular manufacturing chain, which is able to take specific copies [33]. The process of 3D printing needs some power for design slicing before launching the process of manufacturing. 6G technology can accelerate the process by providing a highly efficient computerization model that integrates the efficient supply chain and optimized energy consumption.

1.4.4 Personalized Healthcare

The up-grade of 5G to 6G technology reflects high quality in the patient health-care system. Concurrent monitoring upgrades the preventive and predictive methodology based on intelligent blockchain technology. The personalized e-health approach enables patients to own their data, process the data, and manage their situation based on their environmental condition. Physicians and health care providers can diagnose patients in just minutes, and remedies can be prescribed accordingly. 6G technology integrates augmented and virtual reality devices, the Internet of Things, sensors, and off-body communications [34]. Blockchain technology stores all the real-time communication records in the decentralized mechanism [35]. Such personalized healthcare uses block-chain embedded on 6G networks, which saves patient money and time.

Many commercial implementations have been initiated in health care, such as MedRec [36] and Med chain [37]. Wireless communication–based health-care has inherited visible light communication using visible light as a wave medium with high data access rates. Transmission can be done in a secured manner with low energy consumption and no radiofrequency radiation. Blockchain architecture helps healthcare providers receive real-time data from wireless wearable devices and store it in the block with the help of an immutable distributed ledger. 6G improves the dense communication among massive connections of users using lightweight infrastructure, capable of using the downlink channel.

Soon, with the advancement of flexible fabric telecommunication will lead to transduction of sensations like electrical waves of the human brain to another human brain, which will play the "reality show." 6G technology plays the flexible fabric role for on–off body communication [34], which improves social-psychological scenarios. The adoption of 6G technology in the system will enhance the healthcare model in terms of massive high connectivity of nanoscale devices inside the human body with ultralow latency, making real-time diagnosis easy and capable of operating on high battery capacity. Personalized healthcare will meet tall health parameters in all aspects of new services. Some of the significant challenges solved by blockchain technology in 6G networks for deploying Industrial Application 4.0 are depicted in Figure 1.3.

Scalability: The Industrial Application's primary requirement beyond the system is scalability. But the traditional centralized system is replaced by a decentralized mechanism that increases service network strength [35]. The smart contract design leads to creating a massive number of devices that can communicate with one another on a 6G network and that will meet demand anticipated in the coming era of 2025.

Minimal Latency: Zero delay with sturdy accuracy is the critical demand in future device-to-device communication. Blockchain technology utilizes the 6G frequency for the precise process in deploying Industrial 4.0

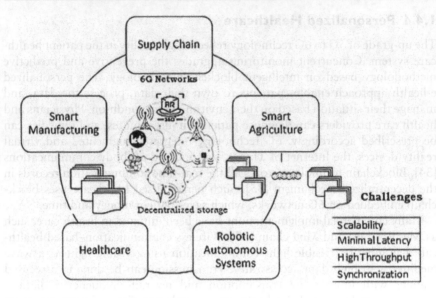

Figure 1.3 Challenges solved by decentralized storage.

applications such as automated driving or wearable devices for medical care with high transmission of data transactions in real-time.

High throughput: The base stations that consist of critically designed systems, connecting millions of devices, should handle the enormous transaction at a given time. This challenging mission is done by blockchain technology; 6G can connect massive devices through the Compass Access System, and everything can connect in a decentralized manner [35], eradicating the single point of failure possibility.

Synchronization: The time-related Industrial 4.0 revolutionary applications, like healthcare wearable devices, need significant synchronization. The automated vehicular designed networks need accuracy without zero delays to accomplish the mission properly in a dynamic environment.

1.5 DISTRIBUTED LEDGER TECHNOLOGY

The stakeholders of multiple cross-industry applications show a keen interest in the 6G ecosystem integrated with distributed decentralized ledger technology. These digitized twins create decentralized configurations in the marketplace with 6G virtualized networks. The business models of 5G technology [38] with mobile network operators operating on centralized mechanisms were replaced by dynamic 6G technology deployed with software-defined networks, network function virtualization, edge service business models, including micro-operator novel services and vertical services in development. The 6G business model developed with blockchain technology enables new use-case ideas on monitoring and protecting the environment, data analytics and data sharing, and intelligent resource allocation.

1.5.1 Monitoring and Protection of the Environment

Along with 6G, global-scale blockchain can act as a decentralized sensing application in the environment. Several notable use-cases are smart cities, smart transportation, smart factory, smart supply chain, smart healthcare. The 6G network architecture, which coordinates sensing and communication, facilitates sharing real-time data in monitoring the environment [35]. Terahertz communication has spectral efficiency with improved wide bandwidth using advanced multiple input/multiple output technology. The critical property for the protection and monitoring of a smart environment includes a very high data rate that depends on wide bandwidth. Terahertz signals work with many antennas and reduce the interference caused in overcoming the range limitations. The excellent features of blockchain deployed with 6G technology will drive smart cities with improved quality of life through intelligent devices. Smart devices that use wireless information energy, as opposed to batteries that need changing, can continuously sense the dynamically changing environment and constantly transfer real-time data.

1.5.2 Data Analytics and Data Sharing

The traditional way of sharing real-time data by adopting machine learning techniques relies on a centralized server mechanism that is highly prone to security, where possibly a single point of failure can occur. Researchers' primary purpose is how machine language will be integrated into the 6G high-speed wireless environment. Data analytics and the processing of data incur significant overhead in centralized ML schemes not suited for 6G networks.

The infrastructure of blockchain technology creates an interface by involving all users in the network in the sharing of shared data. Data analytics thoroughly reconsider reliable real-time data to confirm the decision. Reliable data can be improved by the time-stamping features used in the blockchain, which enhance confidence in the security of data shared on the network and create the trust to make decisions.

The sharing of data reveals that the information moves relentlessly on the 6G ecosystem and that each subsection of data needs authorization from the owners. The standardized smart contract improves the automated sharing of data and is transparent to all the users in the network [11]. With the elaboration of blockchain technology, the 6G network provoked many new emerging infrastructures on integrated terrestrial space and federated networks.

The industrial cyber–physical system, which is decentralized in nature and whose components are distributed physically over all places of a smart factory, will collect, process, and store physical processes. The analysis and processing of data control physical process decisions with real-time data. There are issues associated with the blockchain-based industrial cyber–physical system connecting many IoT devices that operate on batteries [39]. The consensus power consumption algorithm with 6G technology uses wireless information transfer technology to improve the power in IoT batteries.

1.5.3 Standards of Blockchain Adopted for 6G Technology

The blockchain framework is used as the state-of-the-art technology for 6G to manage the spectrum, network monitoring, and resource utilization. The inherent transparency helps in finding the utilization of the range and record in real time, thereby improving the efficiency of the spectrum. Wireless 6G networks consist of many cells and millions of devices to communicate with one another and face scalable issues.

The standards of blockchain technology to eradicate the potential challenges of 6G wireless network communication are discussed next.

1.5.3.1 Intelligent Resource Management

By 2030, colossal connectivity is expected in the wireless ecosystem. Management of the network will be the consummate challenge in the 6G communication network—making the excellent 6G infrastructure compatible with specific operations like spectrum sharing or decentralized mechanism. Orchestration by Mafakheril et al. [40] has used blockchain technology for self-organizing by utilizing smart contract and resource sharing and adopting the decentralized behavior of blockchain technology.

Maksymuyuk et al. [41] applied improved blockchain technology to propose architecture on intelligent resource management that handles the event between the primary users. The operators or users are controlled through smart contracts. Deep reinforcement learning achieves efficient resource management by adopting blockchain technology for services. The spectrum is allocated dynamically by integrating computable resources and tradeable spectrum. Network slicing is maintained by spectrum, allocated dynamically, and virtualized hardware is facilitated to enhance blockchain-enabled efficient, intelligent resource management. The virtualization of hardware relies on the execution of a smart contract, consisting of scripts run on the network node on the virtual stack. The results of performance are stored as a record on the blockchain. The immutability nature of the contract and automation process excludes the dereliction of the agreement.

1.5.3.2 Access Control of Intelligent Resource Management

Reliance on access control mechanisms in the blockchain diverges as a public and private chain. The public chain provides a secured, reliable network by a proof-based consensus protocol for each primary user. It organizes the participants to improve the efficiency of the 6G community. The private chain, based on permission-based control, needs approval for entry into the chain—given the high-level security threats on open-access blockchain technology. Hence a harder consensus mechanism was adopted for preventing external breaches by third parties. The selection of a suitable consensus mechanism acts as a pillar of the 6G network and clinches consistency, integrity of the distributed nodes in scalability, delay, and security. Rodrigues et al. [11] presented an access control mechanism that prevents DDoS attacks on wireless communication networks with blockchain technology and SDN on 6G networks.

1.5.3.3 Self-Adaptive Smart Contract Scheme

The proposed architecture of blockchain in the 6G network will automatically automate data transmission and control data information. The leader selected on the chain establishes the smart contract consisting of data type and content for automating the control on transmission of data. The leader generates the new block, and a smart contract is recorded in the league, which is broadcast to all the nodes in the same network. Then the nodes will compute the consensus mechanism to verify the contract, and the validation of the smart contract is checked to make the node valid. This smart contract in blockchain technology supports the automated process of data transmission, thereby achieving trust among developers.

The blocks store all the critical data and history of transmission. Sometimes, when conditions in the environment call for an update of the smart contract, changes can be made by means of two methods: automatic smart contract update and artificial contract update. In automatic smart contract updates, the network automatically checks the dynamic environment. If it finds variation, then it automatically updates the contract. The contract follows the same underlying concept for transmission of data for updating the agreement. The artificial update is done by the leader elected in a new block who pushes the other block to achieve automatic control over data transmission.

1.6 INTEGRATED NETWORKS

The various combinations of research in information technology and telecommunication lead to the integration of 6G networks and blockchain technology in three major areas: (1) Internet of Things for interoperability in various business applications, (2) artificial intelligence and machine learning for efficient

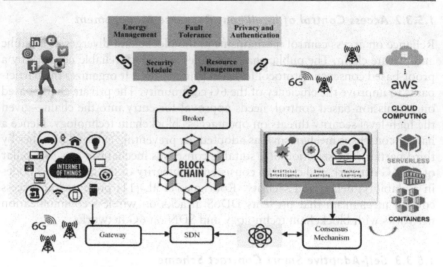

Figure 1.4 Multicross-industry integration of 6G networks and blockchain technology.

energy trading in 6G networks, and (3) dedicated applications such as smart farm, supply chain, tourism, distributed energy system, digital content management, as depicted in Figure 1.4.

1.6.1 Internet of Things for Interoperability in Various Business Applications

The backbone of many business applications in IoT architecture is the evolution of new technologies in functionality, availability, scalability, and maintainability. The main concern of IoT architecture is security for all layers. The issues for most IoT architecture–based devices are due to the network's restricted capacity. The computational power of IoT devices is limited, with low capacity for storing data. Singh et al. [42] identifies many cyber threats on business applications and suggests that blockchain technology can improve the security model on IoT devices.

The mushrooming of 6G/IoT smart applications are anticipated to encourage innovation and bring on a revolutionary ecosystem. The Internet of Everything connects various people with devices in an intelligent way. The consolidation of updated blockchain technology with 6G solves the challenge of interoperability over various business applications. The 6G/IoT network will provide continuous networking service to each participant who faces several issues that blockchain technology can deal with, as follows:

Improvement in 6G/IoT device security: The lightweight, efficient cryptography algorithm guarantees that data is recorded in the ledger by IoT devices over the 6G network, where intruders cannot alter the massive connectivity with high computational power.

Decision on transparency on 6G/ IoT data: Transparency can be permissioned or permission-less for a specific group of people, who can access particular data to improve the confidentiality of data recorded over the blockchain.

Data access on 6G/IoT: Due to the distributed nature and peer-to-peer communication, data access is guaranteed even during disruption of data between the users on the network.

Energy efficiency on 6G/IoT devices: The mining process in blockchain uses different kinds of consensus algorithms like script. XII uses miniblockchain [42] as the alternative, which reduces the computational power nodes, thereby improving the battery life of wearable devices. The lower computational power algorithm is used to improve battery life, thereby lowering the energy on the mining process.

With these efficiency features of blockchain technology, the 6G-based super-IoT network supports symbiotic radio and enhanced IoT-based satellite communications. 6G/IoT symbiotic radio uses licensed bandwidth on dedicated GSM-based spectrum bands that achieve the fastest data transaction between super-IoT devices and access points. The blockchain verification protocol has been securely utilized for spectrum sharing. Terahertz communication and deployment of VLC improve density in the access points of the network. The different types of spectrum sharing supported by blockchain technology were proved by Saad et al. [43] on primary, secondary, and cooperative sharing. The blockchain verification protocol usage on spectrum sharing provides better results than Aloha medium access protocol.

The blooming 6G networks use satellite-based communication to support future terrestrial-based communication. Satellite-based communication provides the global coverage of wearable devices spread out throughout the world. The challenge in satellite-based IoT devices is information from the satellite to IoT device in direct mode. 6G technology, which uses decentralized architecture, can integrate satellite, terrestrial, and airborne. The second challenge is that the distance from IoT devices to the satellites is too large, so the device needs heavy transmitting power. Terahertz communication and optical band expand the spectrum to unlock the broader bandwidth for transmission. Hence every new innovative business model in IoT can be integrated with 6G, and decentralized architecture in various industries is possible.

1.6.2 AI and Machine Learning for Superefficient 6G Networks

The generations that have evolved over time have different expectations of innovations related to various attributes such as latency, data access rate, energy efficiency, spectrum, and coverage. Researchers look to 6G networks with high support quality service and unlimited connectivity for many devices. Beyond the 5G technologies, 6G has been dynamically restructured by integrating artificial intelligence to improve the quality of service.

Future demands on 6G include ultrahigh data rate with low latency and good reliability with energy efficiency on massive connectivity of up to 107 devices/km^2, where every user on wide bands up to 1–3 THz is interlinked with the help of AI. The robust analysis techniques of AI, such as the ability to learn, ability to optimize, and recognition, can be applied on a 6G network. With 6G, networks with decentralized architecture can improve optimization, automate the structure of their own organization of mobile networks, and perform complex decision making on slicing the networks to small-cell infrastructure.

6G infrastructure supports high reliability and leads to accelerated sensing. The sensing on real-time data intelligently collects the heavy dynamic real data and stores it on a decentralized database. The creation of diverse data sets is made possible by using blockchain technology in a suburbanized permissionless network that everyone around the web can view. Coordinating the nursing application programming interface (API) with blockchain API enable communication with the artificial intelligence agents.

Innovative-based AI applications can be designed with emerging algorithms on a complex knowledge set through 6G decentralized networks.

The massive connectivity on 6G networks with their decentralized nature creates the need to make intelligent decisions at various granular levels. Artificial intelligence, with distributed resources such as computation, communication, caching of data, and data control, needs to be trained at edge-based patterns stored in a centralized cloud, which leads to threats to privacy. Using the decentralized nature of blockchain helps keep the records of granular data and variables immutable for computing the data by AI for a better decision-making process. This makes auditing the entire model very easy and creates trust in conclusions made by AI decision-making programs. For example, FINALIZE is the software that combines blockchain technology with machine learning to improve civil-based infrastructure. The tools of FINALIZE automate the construction for speedup in workflow, maintenance, and verification.

Machine learning and artificial intelligence techniques sometimes fail to increase performance, but then some changes in the configuration of the 6G model are needed. Intelligent 6G networks have been classified as operational intelligence, environmental intelligence, service intelligence.

Operational intelligence allocates resources like band and power to achieve better network operations, such as density, scalability, and heterogeneity. Luong et al. [44] identified many issues on caching and data offloading that can be rectified by means of the decentralized nature of blockchain. Environmental intelligence uses an intelligent framework for D2D communication in 6G networks. But it uses edge or cloud devices to store the extracted initial data. So that cryptographic algorithms protect the transmission. Service intelligence includes healthcare, maintenance provided by indoor and outdoor IoT multidevices for 6G intelligence in the human-centric network. The issue in 6G networks of balancing privacy vs. intelligence is significant since it is human-centric. The blockchain technology that eradicates third-party decisions helps store privacy-based medical records in the chain, improving trust and security.

1.6.2.1 6G Intelligent Smart Radio Communication

Smart intelligent radio communication is a prospect for bringing forth highly configured wireless networks with sustained wireless connectivity: a smart radio environment empowered by AI and ML that recycles radio waves [45]. The massive trustless nodes are united to form large-scale authorized, trustworthy networks by means of the unique attributes of blockchain technology. Wireless communication faces some challenges in feasibly interconnecting the physical with a digital platform. The following are some of the challenges in 6G intelligent communication:

Intelligent 6G on network security and privacy: Since intelligent 6G networks act within humanoid networks, algorithms communicate with the private data. The solution is based on a third-party acting as the intermediary between algorithm and primary user. All the private data handled by the anonymized third-party broker is a challenging issue in a super-intelligent 6G network. To solve the problem, innovative technology is essential to maintain privacy. Security and confidentiality encourage trust in blockchain technology, providing the balance in a 6G intelligent network of security and privacy.

Intelligent 6G network security and spectral efficiency: Achieving end-to-end encryption in wireless attack-proof communication requires secured spectrum bandwidth to transmit data. Efficient encryption algorithms imply that a complex computational mechanism will improve bastion security. Consensus mechanisms such as Proof of Work or Proof of Stake are highly complex computational algorithms used in blockchain technology that help in the end-to-end encryption algorithm.

Improve wearable intelligent device battery life: The 6G intelligent network with energy-efficient performance utilizes many micro- and macrosensors in the IoT ecosystem. To minimize energy utilization, the user node computational function can utilize the miniblockchain to improve battery life. 6G communication can use this strategy and will invoke the efficiency of wearable intelligent device communication.

The summarization of AI and ML methods on blockchain-empowered 6G will be needed in new intelligent network architecture for intelligent-based devices, in order to ensure communication with several innovative features.

1.6.3 Dedicated Applications of 6G Networks

Blockchain taxonomy can be deployed into 6G networks for real-time-based applications in agriculture, supply chain logistics, hospitality, tourism centers, and manufacturing [46]. Some of the dedicated applications of 6G networks can use blockchain to develop new technology.

1.6.3.1 Smart Agriculture

The smart farm agriculturally based application critically needs to manage inputs. Caro et al. [47] proposed that an IoT-based solution in agriculture uses centralized server architecture and thus had many threats like data integrity, data tampering, and many single points of failure. He proposed AgriBlockIoT [47] for integrating the massive IoT devices connected in order to automate smart contract implementation, thereby achieving transparency, that is, the storing of the data in the immutable record. The performance of 6G technology that adopts blockchain for smart farms can use AgriBlockIoT to improve architecture in terms of low latency, loading of real-time data, processing and maintaining data, and automatically controlling network traffic.

1.6.3.2 Smart Supply Chain and Logistics

The supply chain and traditional logistics system are employed to store orders and deliver them to their destinations. This old-era system faces shortfalls in several aspects, like transparency, auditability, and accountability. The research and development organization deploys IoT devices on 6G networks with blockchain architecture observed remotely. Sharing data improves trustworthiness for the members of the smart supply chain, as proposed by Weber et al. [48]. The public chain credit evaluation system has helped in the supervision of wireless communication and will improve the supply chain auditing system. Leng et al. [49] identified the double chain agricultural system focusing on storing data and an efficient consensus mechanism that deals with openness and security on the transmission of data. This proposed model may satisfy the smart supply chain objective in terms of cost, quality of service, speed, and dependence on resources, leading to customers' economic satisfaction.

1.6.3.3 Benefits of Blockchain in 6G Networks

The framework of 6G networks, navigated by massive demands in heterogeneous environments, will turn up new business models. The critical role of blockchain technology will take the edge off challenges of 6G networks as follows:

Impact on high security: Blockchain technology uses the various consensus mechanisms on 6G networks and acts as the pillar of the architecture.

Selecting the highly complex consensus mechanism promises consistency and integrity across the globally distributed wireless communication nodes. Mohammed et al. [50] made comparisons on many consensus mechanisms on six aspects such as better latency, consumption of energy, and improved performance on scalability. With this backdrop, the blockchain creates a trusted framework characterized by high security and privacy.

Impact on access: Fairness of access can be achieved with blockchain-based techniques. All transactions are maintained in blocks, as along with the history for each. In the old transaction era, carrier-sensing multiple accesses were not aligned respectively for further access. The intelligent smart contract techniques [11] use a standard threshold for the particular timing when users can gain access for that limited threshold.

Impact on the efficiency of the spectrum: The characteristics of blockchain technology upgrade the approaches of spectrum management through spectrum auction [16, 17]. The hoarding of primary user data in a decentralized manner records idle spectrum data so that secondary data initiates the transaction to access the spectrum by unlicensed users. The blockchain framework excludes fraud users based on transparent behavior.

1.7 CONCLUSION

Blockchain technology unwraps many challenges that lead to a step-up in the benefits of 6G networks. This chapter discusses the prospects of better using blockchain technology for 6G. It also describes blockchain technology standards and supports in solving the issues with future 6G networks on spectrum management, intelligent network allocation, spectrum sharing, energy trading models. The future of dedicated decentralized-based 6G networks was discussed based on their performance and security. The role of blockchain in 6G will concern all the requirements of citizens who can own their data that can be stored and shared securely in this integrated environment with the privacy-preserving nature of blockchain technology and robust security applications.

REFERENCES

[1] Alsharif M, Nordin R (2017). Evolution Towards Fifth-Generation (5G) Wireless Networks: Current Trends and Challenges in Deploying Millimeter Wave, Massive MIMO, and Small Cells. *Telecommun. Syst.*, 64(4):617–637.

[2] Albreem M et al. (2020). A Robust Hybrid Iterative Linear Detector for Massive MIMO Uplink Systems. *Symmetry*, 12(2):306.

[3] Mumtaz S et al (2017). Terahertz Communication for Vehicular Networks. *IEEE Trans. Veh Tech*, 66(7):5617–5625.

[4] Hewa T, Gür G, Kalla A, Ylianttila M (2020). The Role of Blockchain in 6G: Challenges, Opportunities and Research Directions. *2nd 6G Wireless Summit*:1–5.

[5] 1st 6G Wireless Summit (2020). www.6gsummit.com.

[6] Liyanage A, Kumar P, Ylianttila M (2020). *IoT Security: Advances in Authentication.* John Wiley & Sons.

[7] Prados-Garzon J, Adamuz-Hinojosa O, Ameigeiras P et al (2016). Handover Implementation in a 5g SDN Based Mobile Network Architecture. In: *IEEE 27th Annual International Symposium on Personal, Indoor, and Mobile Radio Communications (PIMRC)*. IEEE, pp. 1–6.

[8] Fan K, Ren Y et al (2017). Blockchain-Based Efficient Privacy Preserving and Data Sharing Scheme of Content-Centric Network in 5G. *IET Commun.* (12):527–532.

[9] Yang H, Zheng H et al (2017). *Blockchain Based Trusted Authentication in Cloud Radio Over Fiber Network for 5G.* Paper presented at the 16th International Conference on Optical Communications and Networks, IEEE, pp. 1–3.

[10] Ortega V, Bouchmal F, Monserrat J F (2018). Trusted 5G Vehicular Networks: Blockchains and Content-centric Networking. *IEEE Veh. Technol. Mag.*, 13(2):121–127.

[11] Rodrigues B, Bocek T, Lareida A, Hausheer D et al (2017). A Blockchain-Based Architecture for Collaborative DDoS Mitigation with Smart Contracts. In: *IFIP International Conference on Autonomous Infrastructure, Management and Security*. Springer, pp. 16–29.

[12] Latva-aho M, Leppanen K (2019). Key Drivers and Research Challenges for 6g Ubiquitous Wireless Intelligence. In: *White Paper, 6G Flagship Research Program*. University of Oulu.

[13] Zhang Z et al (2019). 6G Wireless Networks: Vision, Requirements, Architecture, and Key Technologies. *IEEE Veh. Technol. Mag.*, 14(3):28–41.

[14] David K, Berndt H (2018). 6G Vision and Requirements: Is There Any Need for Beyond 5G. *IEEE Veh. Technol. Mag.*, 13(3):72–80.

[15] Shamili P et al (2020). Understanding Concepts of Blockchain Technology for Building the DApps. In: Dash S, Das S, Panigrahi B K (Eds.), *Intelligent Computing and Applications. Advances in Intelligent Systems and Computing*, vol. 1172. Springer. https://doi.org/10.1007/978-981-15-5566-4_33.

[16] Dai H N, Zheng Z, Zhang Y (2019). Blockchain for Internet of Things: A Survey. *IEEE Internet Things J.* doi:10.1109/IoT.2019.2920987.

[17] Sun Y, Zhang L, Feng G et al (2019). Performance Analysis for Blockchain Driven Wireless IoT Systems Based on Tempo-Spatial Model. In: *2019 International Conference on Cyber-Enabled Distributed Computing and Knowledge Discovery*. ICEIS, pp. 348–353.

[18] Li A, Han G, Rodrigues J (2017). Channel Hopping Protocols for Dynamic Spectrum Management in 5G Technology. *IEEE Wirel Commun.*:102–109.

[19] Liang Y C (2020). *Blockchain for Dynamic Spectrum Management, Signals and Communication Technology*. Springer.

[20] Xu H, Klaine P V et al (2020). Blockchain Enabled Resource Management and Sharing for 6G Communications. In: *Digital Communications and Networks*. University of Glasgow.

[21] Chang B, Zhang L (2019). Optimizing Resource Allocation in URLLC for Real-Time Wireless Control Systems. *IEEE Trans. Veh. Tech*:8916–8927.

[22] Samdanis K, Costa-Perez X, Sciancalepore V (2016). From Network Sharing to Multi-Tenancy: The 5G Network Slice Broker. *IEEE Commun. Mag.*:32–39.

[23] Underwood S (2016). Blockchain Beyond Bitcoin. In: *Tech. Rep. 11*. Sutardja Center for Entrepreneurship & Technology Technical Report.

[24] Weiss M B, Werbach K, Sicker D C, Bastidas C E (2019). On the Application of Blockchains to Spectrum Management. *IEEE Trans. Cogn. Commun. N.*, 5(2):193–205.

[25] Backman J, Yrjola S, Valtanen K, Mammela O (2017). Blockchain Network Slice Broker in 5G: Slice Leasing in Factory of the Future Use Case. In: *Joint 13th CTTE and 10th CMI Conference on Internet of Things—Business Models*. Users, and Networks, 2018, January:1–8. https://doi.org/10.1109/CTTE.2017.8260929.

[26] Alladi T, et al (2018). Blockchain Applications for Industry 4.0 and Industrial IoT: A Review. *IEEE Access* (7):176935–176951.

[27] Perboli G, Musso S, Rosano M (2018). Blockchain in Logistics and Supply Chain: A Lean Approach for Designing Real-World Use Cases. *IEEE Access* (6):018–028.

[28] Kapitonov A, Lonshakov S et al (2017). Blockchain Based Protocol of Autonomous Business Activity for Multi-Agent Systems Consisting of UAVs. In: *Proceedings of the Workshop on Research*. Education and Development of Unmanned Aerial Systems Linkoping, 3–5 October.

[29] Decentraland Official Web Site. https://decentraland.org. Accessed 25 February 2019.

[30] VibeHub Official Web Site. www.vibehub.io. Accessed 25 February 2019.

[31] Bitcoin Magazine New (2019). The Lucyd Story: Augmented Eye Ware Backed on the Blockchain. https://bitcoinmagazine.com/articles/lucyd-story-augmented-eyeware-backed-blockchain. Accessed 5 February 2019.

[32] 3D-TOKEN Official Web Site. https://3d-token.com. Accessed 25 February 2019.

[33] Holland M, Stephani's J et al (2018). Intellectual Property Protection of 3D Print Supply Chain with Blockchain Technology. In: *Proceedings of the IEEE International Conference on Engineering*. Technology and Innovation.

[34] Mucchi L et al (2020). How 6G Technology Can Change the Future Wireless Healthcare. In: *2nd 6G Wireless Summit (6G SUMMIT)*. Levi, Finland, pp. 1–6.

[35] Chowdhury M Z, Shahjalal M et al (2019). 6G Wireless Communication Systems: Applications, Requirements, Technologies, Challenges, and Research Directions. arXiv.

[36] MedRec (2016). https://medrec.media.mit.edu. Accessed 12 April 2019.

[37] Medicalchain. (2019). https://medicalchain.com/en. Accessed 12 April 2019.

[38] Sheth K et al (2020). A Taxonomy of AI Techniques for 6G Communication Networks. *Comput. Commun.*, 161:279–303.

[39] Fernández-Caramés T M, Fraga-Lamas P (2019). A Review on the Application of Blockchain to the Next Generation of Cybersecure Industry 4.0 Smart Factories. *IEEE Access*, 7:45201–45218.

[40] Maksymuk T, Gazda J et al (2019). Blockchain-Based Intelligent Network Management for 5G and Beyond. In: *3 International Conference on Advanced Information and Communications Technologies*. IEEE, pp. 36–39.

[41] Mafakheri B, Subramanya T et al (2018). Blockchain Based Infrastructure Sharing in 5G Small Cell Networks. In: *14th International Conference on Network and Service Management*. IEEE, pp. 313–317.

[42] Bruce J D (2019). The Mini-Blockchain Scheme. https: // www.weusecoins.com /assets /pdf /library/The% 20Mini-Blockchain%20Scheme.pdf. Accessed 25 February 2019.

[43] Saad W, Bennis M, Chen M (2019). A Vision of 6G Wireless Systems: Applications, Trends, Technologies, and Open Research Problems. arXiv:1902.10265.

[44] Luong N C, Hoang D T et al (2019). Applications of Deep Reinforcement Learning in Communications and Networking: A Survey. *IEEE Commun. Surv. Tutor*, 21:3133–3174.

[45] Renzo M D, Debbah M et al (2019). Smart Radio Environments Empowered by Reconfigurable AI Meta-Surfaces: An Idea Whose Time Has Come. *J Wireless Com Network*:129.

[46] Singh J, Michels J D (2018). Blockchain as a Service (BaaS): Providers and Trust. In: *Proceedings—In 3rd IEEE European Symposium on Security and Privacy Workshops, EURO S and PW 2018*. Institute of Electrical and Electronics Engineers Inc., pp. 67–74.

[47] Caro M P, Ali M S et al (2018). Blockchain Based Traceability in Agri-Food Supply Chain Management: A Practical Implementation. In: *Proceedings IoT Vertical and Topical Summit on Agriculture*-Tuscany (IoT Tuscany), pp. 1–4.

[48] Weber I, Xu X et al (2016). Untrusted Business Process Monitoring and Execution Using Blockchain. In: *Proceedings of International Conference Bus*. Process Manage, pp. 329–347.

[49] Leng K et al (2018). Research on agricultural supply chain systems with double chain architecture based on blockchain technology. *Future Gener. Comput. Syst.*, 86:641–649.

[50] Mohammed H et al (2020). Research Activities, Challenges and Potential Solutions Based on Blockchain Technology. In: *Symmetry*. MDPI.

Chapter 2

Empowering 6G Networks through Blockchain Architecture

Kaustubh Ranjan Singh[1] and Parul Garg[2]

CONTENTS

2.1 INTRODUCTION

Wireless connectivity is now a ubiquitous utility, which now significantly links the various aspects of human life, such as transport, healthcare, entertainment, etc. 6G communications present several advantages compared with previous generations of cellular technology, as is evident in Table 2.1. Providing enhanced user experience requires overhauling the network infrastructure to meet the unprecedented service level and quality of service requirements for various applications like augmented reality, machine-to-machine communication,

[1] Department of Electronics and Communication Engineering, Delhi Technological University, Delhi, India
[2] Department of Electronics and Communication Engineering, Netaji Subhas University of Technology, Delhi, India
Corresponding author: kaustubh@dtu.ac.in

DOI: 10.1201/9781003264392-2

Table 2.1 Comparison of Key Features in 4G, 5G, and 6G

	4G	5G	6G
Data rate (Gbps)	1	10	1000
Latency (ms)	100	10	1
Spectral efficiency (bps/Hz)	15	30	100
Mobility (km/h)	Up to 350	Up to 500	Up to 1000
Self-driving vehicle	No	Limited	Yes
Augmented reality	No	Limited	Yes
Haptic communication	No	Limited	Yes

and holographic communication. The proposed next-generation 6G network [1–3] infrastructure envisions the network elements to be virtualized, softwarized, and cloudified, promoting the development of new network services and real-time applications. This will also provide hyperconnectivity to heterogeneous device ecosystems comprised of the Internet of Things (IoT), machine-to-machine (M2M), device-to-device (D2D) spread across ground, underwater, air, and space.

2.2 TRENDS FOR 6G

1. *Data rate and device density*: 6G networks will be required to cater to the huge surge in mobile data traffic, which is projected to exceed 5016 EB/month by 2030 [4, 5], as shown in Figure 2.1. The high density of interconnections (1 million/km^2) projected for the 6G network needs optimum resource utilization and requires exploring high-frequency bands 73–140 GHz and 1–3 THz for providing highly reliable user-centric scenarios with 99.99% reliability at 1 ms user plane latency.
2. *Spectrum and energy efficiency*: 6G will also provide 3D-spatial connectivity for aerial users, UAVs, drones, and thus the spectral [6] and energy efficiency requirements of different scenarios need to be redefined by using volumetric definitions of bps/Hz/m3/joules, evolving from 2G (bps) to 5G (bps/Hz/m3/joules) [6, 7].
3. *Emergence of smart surfaces and meta-lens for beamforming*: The directed beamforming for users is phase-shifted through a metasurface lens, which is applied on the signal [8] with an antenna array. DC bias is applied with this beam, which helps to adjust beam direction as well, as shown in Figure 2.2, and which sharpens the beam shape and directivity, leading to improved received power [9–11]. In cases where no direct propagation path exists between the receiver and transmitter,

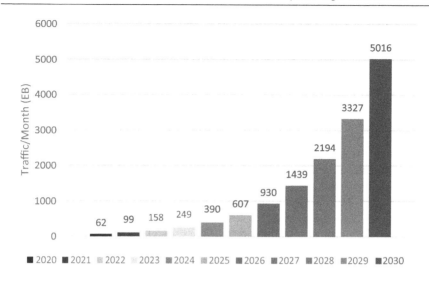

Figure 2.1 Global mobile data traffic in 2020–2030 forecast by ITU.

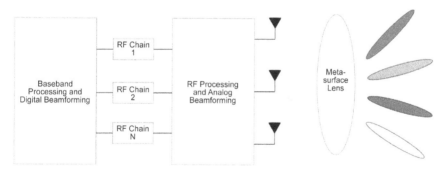

Figure 2.2 Metasurface lens for beamforming.

reconfigurable intelligent surface (RIS) can also be used for the wireless communications link. Recently proposed RIS integrated with metamaterial surfaces [12] provides effective performance and have been deployed on roads, walls, and sometimes entire buildings. This also has tremendous potential to drive 6G architectural evolution for enhancing wide coverage to connected devices.

4. *Self-organizing networks (SON)*: SON technology stimulates the need for providing intelligent solutions to effectively manage aspects of a network-like operations, resources, and optimization into a self-sustaining network (SSN) with improved key performance indicators (KPIs).

2.3 SERVICE CLASSES FOR 6G NETWORKS

The following service classes have been envisioned supporting various applica-tion and device scenarios, as shown in Figure 2.3:

1. *Three-dimensional integrated communications (3D-InteCom)*: Since 6G involves aerial communications through UAVs and underwater commu-nications, there is a need to incorporate three-dimensional volumetric scenarios for network planning, analysis, and optimizations, which is a significant departure from the 2D real analysis for network planning done hitherto.
2. *Secure ultrareliable low-latency communications (SURLLC)*: This pro-vides network services with low latency (1 ms) and very high reliability (99.999%) for applications that are of critical nature like smart grid and remote surgery.
3. *Enhanced mobile broadband (eMBB) Plus*: This provides a high quality of experience with greater data bandwidth, complemented by latency improvements for facilitating design of big data–driven high-capacity communication services without the issues of secrecy and user privacy.
4. *Unconventional data communications (UCDC)*: This affects open-ended and novel application scenarios that involves holographic, human, and tactile communication. Holographic communication enables the real-time projection of objects, people, and environment with realism that rivals physical presence. Human bond communication incorporates olfactory and tactile elements that allow more expressive and holistic sensory information exchange through communication techniques for more human sentiment–centric communication.
5. *Big communications (BigCom)*: BigCom in 6G will provide high-speed throughput for large urban and remote area coverage due to

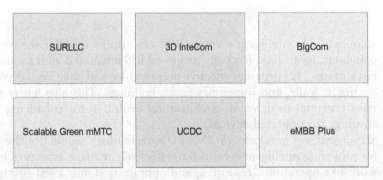

Figure 2.3 6G service classes.

AI-enabled high bandwidth terahertz communication and efficient resource allocation.

6. *Green massive machine-type communication (mMTC)*: This refers the efforts to build unified and ubiquitous network coverage throughout all space dimensions, including volumetric density of devices having an efficient low-energy profile. The energy efficiency can be increased by incorporating the energy harvesting architectures from various sources such as solar, wind, and ambient RF waves.

2.4 OPEN ISSUES FOR BLOCKCHAIN-BASED 6G NETWORKS

Despite the potential advantages offered by 6G, some of the challenges that need to be addressed are as follows:

1. *High propagation loss and beam management for THz frequencies:* The submillimeter wave THz frequencies designated for use in 6G have the drawback of high attenuation [2] by path loss and atmospheric absorption. The atmospheric conditions make its channel modeling [13, 14] even difficult due to absorption and dispersion effects. This will require very careful designing of receiver systems with smart beamforming management [15] to address the very small gain and effective area [16] of different THz band antennas. The RMS delay spread and angle spread for the THz channel model [14] are defined as:

$$^S RMS = \sqrt{\frac{\sum_i \left(d_i - \bar{d}\right)^2 pi}{2}} \tag{2.1}$$

$$^a RMS = \sqrt{\frac{\sum_i \left(a_i - \bar{a}\, mod\, 180\right)^2 pi}{\sum_i pi}} \tag{2.2}$$

where d_i, p_i are delay and received power of path, $\bar{a}$ is the mean angle, and $\bar{d}$ is the mean delay.

The RMS delay spread and RMS angle spread are key components of the study to understand channel correlation and received power from a multipath propagation channel evaluated as shown in Figures 2.4 and 2.5.

2. *Scalability, throughput, and Latency*: With the emergence of massive machine-type communications (mMTC) [17, 18], it will be a big

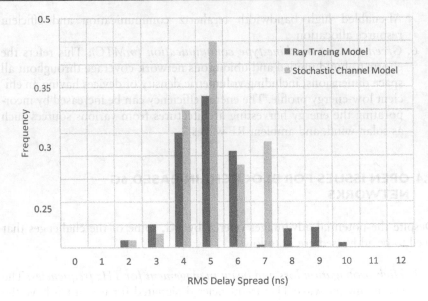

Figure 2.4 **RMS delay spread modelling for THz communication [14].**

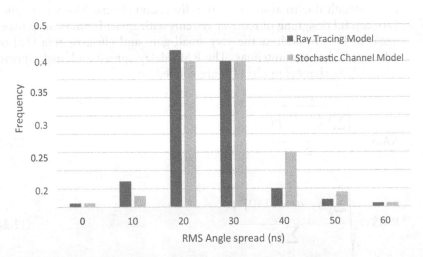

Figure 2.5 **RMS delay spread modeling for THz communication [14].**

challenge to provide a fully scalable solution catering to billions of interconnected devices with high data rates [19, 20] and low latency, especially for mission-critical scenarios like autonomous driving and healthcare systems that involve large volumes of sensory data exchange.

3. *Spectrum management*: For QoS maximization, efficient spectrum management policy [21–23] is necessary to get optimum resource utilization. Concerns regarding spectrum sharing in heterogeneous network and device scenarios [24–26] have to be addressed, along with interference management using successive interference cancellation and parallel interference cancellation.

4. *Privacy and data confidentiality for open networks*: The huge volume of data collected for providing network services [5] needs to be regulated and access given only to the authorized entities, requiring careful balancing of privacy and network intelligence.

5. *High capacity backhaul*: With such high-density access networks having high data rates in 6G, it is necessary to design backhaul networks that can cater to the high volume of data traffic between the core and access network. Optical fiber networks, free space optical networks, offer potential solutions for the backhaul network.

6. *Network availability*: The 6G ecosystem diversification with differentiated virtual networks and high number of interconnected increases the risk of DDoS attacks. AI and quantum-based encryption schemes, besides a secure physical layer, would be required to provide the network resiliency against such attacks.

7. *Storage*: Each participating node has to keep a copy of the transactions, which can become challenging for resource-constrained devices used in D2D, IoT.

8. *High blockchain transactions delay*: Sometimes, there is issue of large latency due to transactions pending for approval and appended into the blockchain network due to block size limitations. This will lead to large delays and may prove challenging for strict QoS applications.

2.5 BLOCKCHAIN FUNDAMENTALS

Blockchain is a peer-to-peer distributed technology that securely and immutably records transactions between the network parties in blocks [27]. In its essence, blockchain provides a trusted, accountable, and transparent environment where intermediaries who were required historically to validate and record transactions are no longer needed. The main features of blockchain are shown in Figure 2.6.

Blockchain, through its cryptographic hashing process, enables timestamped data transactions to become tamperproof, as shown in Figure 2.7. After verification from the blockchain network, the transactions become integrated into the metadata block using a previous hash. A chronological chain of hashed data blocks is created this way, making the chain of data blocks immutable. Another important aspect of blockchain is its decentralized nature; i.e., the validation of transactions is done through consensus protocols rather than a centralized authority. All information about the transactions is recorded and is visible to the network participants, which can be used for various 6G network applications like slicing and resource sharing, where transparent ledgers can support open and secure data solutions.

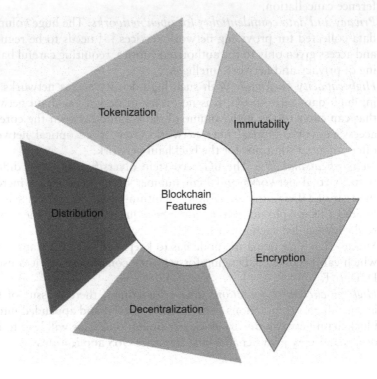

Figure 2.6 Main features in a blockchain network.

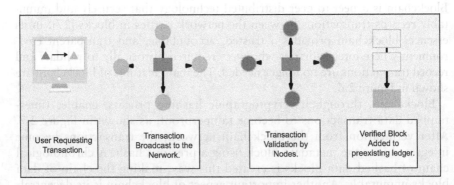

Figure 2.7 Working mechanism of a blockchain.

2.5.1 Components of Blockchain

1. *Data block*: The blockchain metadata has a series of blocks originating from a genesis block that contains a record of transactions that is linked with the previous blocks through the hash label, enabling tracing of all blocks. The structure contains a blockchain header and a Merkel tree-based transaction record. The blockchain header contains the block hash value, the Merkle root for storing transactions, timestamp, and consensus-based nonce value to produce a hash value of target difficulty, as illustrated in Figure 2.8.

2. *Distributed ledger (database)*: It is a shared database, openly accessible to blockchain network users, that records transactions like the data exchange process among the users. The network users make use of consensus-based methods to perform the transaction, without the involvement of any external third party.

3. *Consensus algorithms*: To overcome issues like double-spending in the network, the blockchain relies on consensus algorithms that are a set of protocols used for validation of transactions in the blockchain network. It involves reaching an agreement on a block over unreliable nodes using consensus, such as proof of Work (PoW) algorithm used in Bitcoin to enable miners to perform transactions in an unreliable network. The mining involves performing mathematical computation tasks in the form of a puzzle that uses the computing resources of miners. The

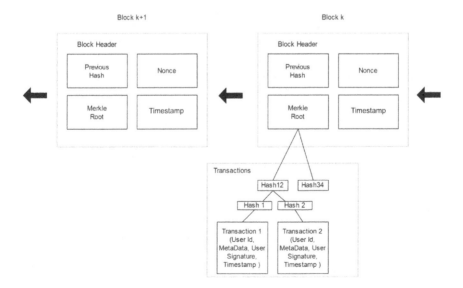

Figure 2.8 Encryption and block generation in the blockchain network.

main drawback of the proof of work algorithm is the high computation requirement, making way for more efficient consensus algorithms like Byzantine fault tolerant (BFT) and proof of stake (PoS).

4. *Smart contracts*: Smart contracts are self-executing applications having operational rules and penalties that run on the blockchain network and perform tasks once the conditions mentioned in the contract clauses are met. Smart contracts allow transactions without requiring any intermediary, unlike traditional systems where a central authority enforces contract conditions.

2.6 BLOCKCHAIN-BASED USE-CASE SCENARIOS FOR 6G

2.6.1 Blockchain for Software-Defined Networking (SDN)

SDN promotes flexibility and programmability of intelligent network architecture and will be a key component of 6G networks. SDN relies on the separation of the control plane and data plane through software provisioning. This will enable access between heterogeneous networks through a logical software controller, providing better user services with efficient network resource use. SDN enables a high degree of softwarization of 6G networks, enabling dynamic traffic flow as shown in Figure 2.9. Despite its advantages, some issues of flexibility, scalability, and privacy continue to pose a challenge for its adoption in terms of security and network management. The separation of data plane and control plane in SDN can increase vulnerability and attack surface, especially during authentication in the control plane and application layer. The SDN controller's centralized design makes it open to attacks in the control layer, causing routers and switches to be compromised. Building a scalable SDN distributed network comprising multiple SDN controllers that communicate by exchanging secure information while offering different services to users remains a challenge.

The centralized network design of current SDN models makes disruption of the entire network a high possibility in the event that a network element is attacked. Thus there is a need to evolve a decentralized SDN architecture that can overcome the possibility of such attacks and provide redundancy and resilience to the overall network architecture in the service provisioning latency issues between different service providers. A blockchain-oriented SDN architecture solution is being proposed to overcome these shortcomings. Blockchains can address potential network and security issues caused by the centralized SDN framework. A decentralized blockchain-based SDN architecture solutions are being worked out to overcome the shortcomings of next-generation 6G networks. The work [28] discusses the role of blockchain technology integration with fault-tolerant control SDN architectures,

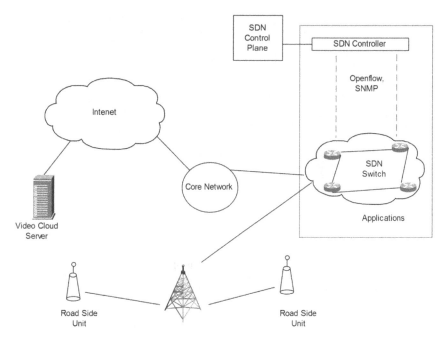

Figure 2.9 Blockchain-based SDN architecture.

as depicted in Figure. 2.10. Using OpenFlow protocol, the data-forwarding functionality is provided in the data plane, software-controlled. Whereas in the control plane, blockchain links all SDN controllers within their respective control domains in a distributed manner.

Figure 2.11 depicts the recovery latency of block Control based on different traffic loads. It results from failure recovery based on blockchain ledgers and smart contracts, thereby reducing the failure recovery latency.

Every SDN controller contains the distributed ledger kept through the consensus

plane. To provide the customized network function, the information from the distributed ledger is utilized through smart contracts. The consensus plane performs the multi-SDN controller consensus operation, which updates the information about processes and services on a distributed data ledger. The contract plane consists of smart contracts that are used to perform different network functions. Such architecture overcomes the issues of fault tolerance and data consistency by using blockchain consensus and distributed ledger system.

Another potential area of SDN-blockchain-based security framework [29–31] is in vehicular ad hoc networks (VANETs). The global functions like

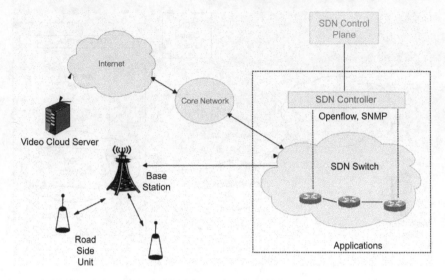

Figure 2.10 Blockchain-based SDN for vehicular ad hoc networks (VANETs).

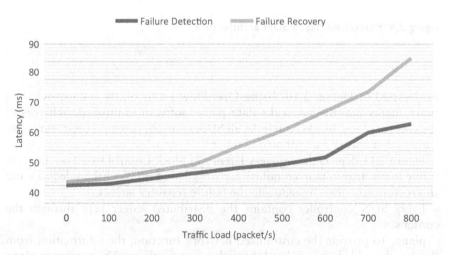

Figure 2.11 Failure detection and recovery of block control [28].

authentication and traffic management are done by the SDN controller and the blockchain through its data ledger, which records all vehicular messages and ensures reliable message transmission through the data plane. The blockchain-SDN-based vehicular system presents a solution for solving potential attacks like denial-of-service (DoS), SDN controller attacks, and unauthenticated access control.

2.6.2 Blockchain for Network Functions Virtualization (NFV)

Network functions virtualization (NFV) is a network architecture concept that uses standard hardware and allows hosting-independent software components for different networks [32]. Standardized by European Telecommunications Standards Institute (ETSI), NFV architecture consists of network function virtualization infrastructure (NFVI), virtualized network functions (VNFs), and management and network orchestration (MANO). VNFs are functions that are executed on NFVI, whereas MANO covers life cycle and management [33] and orchestration of physical and software resources. NFV provides flexible virtualized components and functionalities of the networks by decoupling hardware like firewall and gateways from the functions that run on them to provide the virtualized instances of firewalls and gateways, as shown in Figure 2.12, resulting in a greater reduction in operational and capital expenditure for telecom operators.

While NFVs optimize resources required by VNFs for energy use and cost, they also incur security challenges, especially in scenarios when the end-to-end service provisioning may involve several heterogeneous cloud entities, leading to concerns about data security. Also, the security of virtualized servers running on virtual machines (VM) can be at risk from different horizons. An external denial-of-service attack can be orchestrated by an attacker, which creates a VM that can run in a virtualized server, impacting the service and compromising data integrity. Blockchain technology, with its distributed ledger system, offers solutions to address these potential security risks. Secure orchestration of various VNF services like network management, data auditing, and ensuring system state integrity against both insider attacks and external threats, i.e., malicious VM modifications and DoS attacks, are performed through blockchain.

In Rebello et al. [34], a blockchain-based system architecture for the management of service functions in open platform for network function virtualization (OPNFV) is discussed. The system architecture shown in Figure 2.9 has three modules: the visualization module, the orchestration module, and the blockchain module. After the interface between the tenants and NFV is set up by the visualization module, the orchestration module executes the tenant instructions through the visualization module. The blockchain module performs verification of these transactions before the orchestration module implements them; the blockchain growth rate for the architecture discussed is shown in Figure 2.13.

To deal with the challenge of finding and selecting host VNF infrastructure by the users, the work [35] introduces a blockchain-based solution called BRAIN, a blockchain-based reverse auction solution for infrastructure supply. BRAIN uses smart contracts to achieve trust-based agreements between users and network operators regarding service configurations required and resources contracted for them.

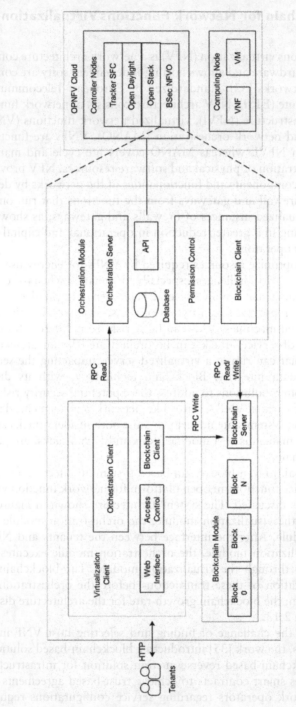

Figure 2.12 Blockchain-based NFV architecture.

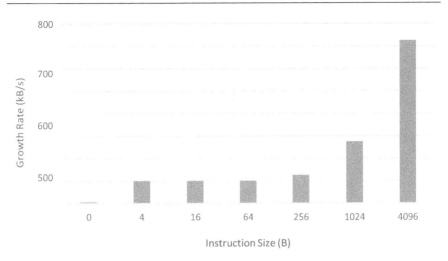

Figure 2.13 BSec-NFVO prototype performance evaluation for blockchain growth rate [34].

2.6.3 Blockchain for Mobile Edge Computing (MEC)

Mobile edge computing (MEC), also known as mobile cloud or fog computing, provides a computing ecosystem for storage, high-processing capabilities, etc. Compared to the remote cloud, edge computing offers low latency and faster service response. Its network elements like servers are positioned at the network edge, integrated with various hybrid networks like cellular, IoT, D2D, M2M. Its distributed architecture allows scalability and flexible network complexity management to cope with the expected growth in demand for the 6G scenario.

However, the security for the MEC architecture is vulnerable to various threats like jamming, denial of service, sniffer attacks, etc., during information exchange from edge servers to the device ecosystem. Owing to a high degree of dynamism in the MEC architecture, ensuring safety from external disruptions and modifications is also of primary concern. Blockchain's decentralized architecture also offers solutions to address these concerns of security and network management in networking, storage, and computation, as shown in Figure 2.14.

A secure authentication system for information exchange between edge servers and different device ecosystems like IoT, D2D can be reliably built using blockchain architecture. The data required for authentication and user access information system can be saved on blockchain ledgers, which can also log mobile terminals' activities. Similarly, the blockchain architecture

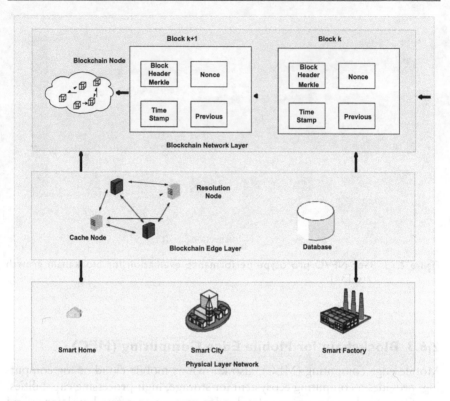

Figure 2.14 Blockchain-based MEC architecture.

can be deployed in vehicular edge computing scenarios to provide secure, low-latency-based data processing services against external malicious attacks.

Further using smart contracts, blockchain provides an access control scheme for enforcing key management and authentication protocol in energy and information flow in vehicular networking without the need for complex cryptographic primitives. The profile management of car, driver, and vehicular data about the car like vehicular sensor data integrated into the MEC network can be kept using blockchain networks. Thus by using blockchain functionality resource management [36], data sharing [37], or resource allocation [38], the performance of edge computing is improved while guaranteeing security properties of the network. Besides, for MEC- based vehicular networks, blockchain data ledger can also be used for decentralized big data repository platforms like InterPlanetary File System (IPFS) [39], Filecoin, and Storij.

2.6.4 Blockchain for Network Slicing

Network slicing, a cellular technology feature evolved in 3GPP's Release-16, refers to specific instances of virtualization allowing logical networks to be operated over shared physical network infrastructure. A network slice consists of several VNFs clubbed together based on specific service class requirements like mMTC, eMBB, and uRLLC, catering to various network vertical services requirements. Multiple virtual networks can be instantiated from a network slice of a single physical network through mobile virtual network operators (MVNOs), which can be orchestrated to accommodate applications highly secure like V2X communications and remote robotic surgery.

Network slicing allows physical hardware to be restricted for building multiple virtual networks by telecom providers to provide specific application services like smart home, vehicular network, and smart factory. Due to open cloud-based network slice architecture instances, malicious attackers can abuse the resources of elastic slice capacity to target other slices and render them out of service through different attack vectors.

However, blockchains can offer significant opportunities for enhancing security in network slice management [40]. Here, manufacturing equipment leases the network slice necessary for independent operations, besides approving the approved service-level agreement (SLA), as shown in Figure 2.15. Blockchain is used in network slice trading [41], whereas smart contracts execute the task of slice operation from the network slice broker.

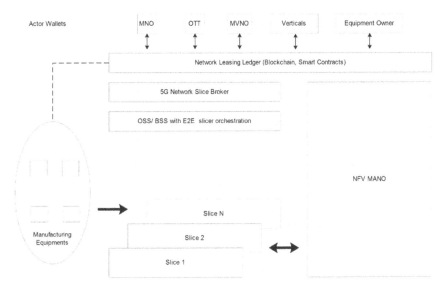

Figure 2.15 Blockchain-based network sliced architecture.

2.7 CONCLUSION

Blockchain has emerged as a potential technology enabler for 6G networks due to its widely acknowledged role in security enhancement and data management. In this chapter, different possible use-case scenarios for providing blockchain-based applications, like SDN, NFV, MEC, etc. have been deliberated, along with their inherent limitations in adopting them.

REFERENCES

1. B. Shihada, M. S. Alouini, S. Dang, and O. Amin. What should 6g be? *Nature Electronics*, 61(3):20–29, April 2020.
2. E. Calvanese Strinati, S. Barbarossa, J. L. Gonzalez-Jimenez, D. Ktenas, N. Cassiau, L. Maret, and C. Dehos. 6g: The next frontier: From holographic messaging to artificial intelligence using subterahertz and visible light communication. *IEEE Vehicular Technology Magazine*, 14(3):42–50, 2019.
3. N. H. Mahmood, H. Alves, O. A. Lo'pez, M. Shehab, D. P. M. Osorio, and M. Latva-Aho. Six key features of machine type communication in 6g. In *2020 2nd 6G Wireless Summit (6G SUMMIT)*, pages 1–5, 2020.
4. G. Liu, Y. Huang, N. Li, J. Dong, J. Jin, Q. Wang, and N. Li. Vision, requirements and network architecture of 6g mobile network beyond 2030. *China Communications*, 17(9):92–104, 2020.
5. T. Nguyen, N. Tran, L. Loven, J. Partala, M. Kechadi, and S. Pirttikangas. Privacy-aware blockchain innovation for 6g: Challenges and opportunities. In *2020 2nd 6G Wireless Summit (6G SUMMIT)*, pages 1–5, 2020.
6. E. Basar. Reconfigurable intelligent surface-based index modulation: A new beyond mimo paradigm for 6g. *IEEE Transactions on Communications*, 68(5):3187–3196, 2020.
7. J. Liu, W. Liu, X. Hou, Y. Kishiyama, L. Chen, and T. Asai. Non-orthogonal waveform (now) for 5g evolution and 6g. In *2020 IEEE 31st Annual International Symposium on Personal, Indoor and Mobile Radio Communications*, pages 1–6, 2020.
8. S. Zeng, H. Zhang, B. Di, Z. Han, and L. Song. Reconfigurable intelligent surface (RIS) assisted wireless coverage extension: Ris orientation and location optimization. *IEEE Communications Letters*, pages 1–1, 2020.
9. E. Basar, M. Di Renzo, J. De Rosny, M. Debbah, M. Alouini, and R. Zhang. Wireless communications through reconfigurable intelligent surfaces. *IEEE Access*, 7:116753–116773, 2019.
10. A. Tarable, F. Malandrino, L. Dossi, R. Nebuloni, G. Virone, and A. Nordio. Meta-surface optimization in 6G sub-THz communications. In *2020 IEEE International Conference on Communications Workshops (ICC Workshops)*, pages 1–6, 2020.
11. M. Di Renzo, A. Zappone, M. Debbah, M. S. Alouini, C. Yuen, J. de Rosny, and S. Tretyakov. Smart radio environments empowered by reconfigurable intelligent surfaces: How it works, state of research, and the road ahead. *IEEE Journal on Selected Areas in Communications*, 38(11):2450–2525, 2020.

12. E. Basar. Reconfigurable intelligent surface-based index modulation: A new beyond mimo paradigm for 6g. *IEEE Transactions on Communications*, 68(5):3187–3196, 2020.

13. S. Priebe, and T. Kurner. Stochastic modeling of the indoor radio channels. *IEEE Transactions on Wireless Communications*, 12(9):4445–4455, 2013.

14. B. Peng, and T. Ku¨rner. A stochastic channel model for future wireless THz data centers. In *2015 International Symposium on Wireless Communication Systems (ISWCS)*, pages 741–745, 2015.

15. A. Gureev, M. Cherniakov, E. Marchetti, and I. Gureev. Channel description in the low-THz wireless communications. In *2017 IEEE Conference of Russian Young Researchers in Electrical and Electronic Engineering (EIConRus)*, pages 1240–1243, 2017.

16. M. Pengnoo, M. T. Barros, L. Wuttisittikulkij, B. Butler, A. Davy, and S. Balasubramaniam. Digital twin for metasurface reflector management in 6g terahertz communications. *IEEE Access*, 8:114580–114596, 2020.

17. S. R. Pokhrel, J. Ding, J. Park, O. S. Park, and J. Choi. Towards enabling critical MMTC: A review of URLLC within MMTC. *IEEE Access*, 8:131796–131813, 2020.

18. K. Mikhaylov, V. Petrov, R. Gupta, M. A. Lema, O. Galinina, S. Andreev, Y. Koucheryavy, M. Valkama, A. Pouttu, and M. Dohler. Energy efficiency of multi-radio massive machine-type communication (MR-MMTC): Applications, challenges, and solutions. *IEEE Communications Magazine*, 57(6):100–106, 2019.

19. S. Bo¨cker, C. Arendt, P. Jo¨rke, and C. Wietfeld. Lpwan in the context of 5G: Capability of lorawan to contribute to MMTC. In *2019 IEEE 5th World Forum on Internet of Things (WF- IoT)*, pages 737–742, 2019.

20. E. C. Liou, and S. C. Cheng. A QoS benchmark system for telemedicine communication over 5G URLLC and MMTC scenarios. In *2020 IEEE 2nd Eurasia Conference on Biomedical Engineering, Healthcare and Sustainability (ECBIOS)*, pages 24–26, 2020.

21. M. Matinmikko-Blue, S. Yrjo¨la¨, and P. Ahokangas. Spectrum management in the 6g era: The role of regulation and spectrum sharing. In *2020 2nd 6G Wireless Summit (6G SUMMIT)*, pages 1–5, 2020.

22. L. Lopez-Lopez, M. Matinmikko-Blue, M. Cardenas-Juarez, E. Stevens-Navarro, R. Aguilar- Gonzalez, and M. Katz. Spectrum challenges for beyond 5g: The case of Mexico. In *2020 2nd 6G Wireless Summit (6G SUMMIT)*, pages 1–5, 2020.

23. X. Liu, H. Ding, and S. Hu. Uplink resource allocation for NOMA-based hybrid spectrum access in 6G-enabled cognitive internet of things. *IEEE Internet of Things Journal*:1–1, 2020.

24. R. K. Saha. Licensed countrywide full-spectrum allocation: A new paradigm for millimeter- wave mobile systems in 5g/6g era. *IEEE Access*, 8:166612–166629, 2020.

25. M. Lin, and Y. Zhao. Artificial intelligence-empowered resource management for future wire- less communications: A survey. *China Communications*, 17(3):58–77, 2020.

26. V. Ziegler, H. Viswanathan, H. Flinck, M. Hoffmann, V. Ra¨isa¨nen, and K. Ha¨to¨nen. 6g architecture to connect the worlds. *IEEE Access*, 8:173508–173520, 2020.

27. S. Nakamoto. *Bitcoin: A peer-to-peer electronic cash system*. Satoshi Nakamoto Institute, 2009.

28. H. Yang, Y. Liang, Q. Yao, S. Guo, A. Yu, and J. Zhang. Blockchain-based secure distributed control for software defined optical networking. *China Communications*, 16(6):42–54, 2019.

29. L. Xie, Y. Ding, H. Yang, and X. Wang. Blockchain-based secure and trustworthy internet of things in SDN-enabled 5g-vanets. *IEEE Access*, 7:56656–56666, 2019.

30. D. Zhang, F. R. Yu, and R. Yang. Blockchain-based distributed software-defined vehicular networks: A dueling deep Q -learning approach. *IEEE Transactions on Cognitive Communications and Networking*, 5(4):1086–1100, 2019.

31. N. Kumar, R. Chaudhry, O. Kaiwartya, N. Kumar, and S. H. Ahmed. Green computing in software defined social internet of vehicles. *IEEE Transactions on Intelligent Transportation Systems*, pages 1–10, 2020.

32. F. Z. Yousaf, M. Bredel, S. Schaller, and F. Schneider. NFV and SDN—key technology enablers for 5g networks. *IEEE Journal on Selected Areas in Communications*, 35(11):2468–2478, 2017.

33. R. Mijumbi, J. Serrat, J. Gorricho, N. Bouten, F. De Turck, and R. Boutaba. Network function virtualization: State-of-the-art and research challenges. *IEEE Communications Surveys Tutorials*, 18(1):236–262, 2016.

34. G. A. F. Rebello, I. D. Alvarenga, I. J. Sanz, and O. C. M. B. Duarte. Bsec-nfvo: A blockchain- based security for network function virtualization orchestration. In *ICC 2019–2019 IEEE International Conference on Communications (ICC)*, pages 1–6, 2019.

35. M. F. Franco, E. J. Scheid, L. Z. Granville, and B. Stiller. Brain: Blockchain-based reverse auction for infrastructure supply in virtual network functions-as-a -service. In *2019 IFIP Networking Conference (IFIP Networking)*, pages 1–9, 2019.

36. G. Qiao, S. Leng, H. Chai, A. Asadi, and Y. Zhang. Blockchain empowered resource trading in mobile edge computing and networks. In *ICC 2019–2019 IEEE International Conference on Communications (ICC)*, pages 1–6, 2019.

37. X. Zheng, R. R. Mukkamala, R. Vatrapu, and J. Ordieres-Mere. Blockchain-based personal health data sharing system using cloud storage. In *2018 IEEE 20th International Conference on e-Health Networking, Applications and Services (Healthcom)*, pages 1–6, 2018.

38. C. Xia, H. Chen, X. Liu, J. Wu, and L. Chen. Etra: Efficient three-stage resource allocation auction for mobile blockchain in edge computing. In *2018 IEEE 24th International Conference on Parallel and Distributed Systems (ICPADS)*, pages 701–705, 2018.

39. M. S. Ali, M. Vecchio, M. Pincheira, K. Dolui, F. Antonelli, and M. H. Rehmani. Applications of blockchains in the internet of things: A comprehensive survey. *IEEE Communications Surveys Tutorials*, 21(2):1676–1717, 2019.

40. K. Valtanen, J. Backman, and S. Yrjo¨la¨. Creating value through blockchain powered resource configurations: Analysis of 5g network slice brokering case. In *2018 IEEE Wireless Communications and Networking Conference Workshops (WCNCW)*, pages 185–190, 2018.

41. K. Valtanen, J. Backman, and S. Yrjo¨la¨. Creating value through blockchain powered resource configurations: Analysis of 5g network slice brokering case. In *2018 IEEE Wireless Communications and Networking Conference Workshops (WCNCW)*, pages 185–190, 2018.

Chapter 3

Blockchain Technology and 6G
Trends, Challenges, and Research Directions

Shyam Mohan,[1] S. Ramamoorthy,[2]
Vedantham Hanumath Sreeman,[3]
Venkata Chakradhar Vanam,[4] and
Kota Harsha Surya Abhishek[5]

CONTENTS

[1] Faculty in the Department of CSE, Sri Chandrasekarendra Saraswathi Viswa Maha Vidyalaya (SCSVMV), Enathur, India
[2] Faculty in the Department of CSE, SRM Institute of Science and Technology, Kattankulathur, Chennai, Tamilnadu, India
[3,4,5] Final Year CSE, Sri Chandrasekarendra Saraswathi Viswa Maha Vidyalaya (SCSVMV), Enathur, India
Corresponding author: jsshyammohan@kanchiuniv.ac.in

DOI: 10.1201/9781003264392-3

3.1 INTRODUCTION

In a data-intensive and connected society, all the systems are associated with consistent information with a wide range of remote systems in the air, underwater, and in space [1]. Data is moving fast, and complete automation is possible if 6G networks are quick in providing efficient and quick access [2]. By

2025, 6G data is expected to grow by 607 exabytes/month and by 5016 exabytes/month by 2030 [3].6G applications are given here [4–9]:

1. Extended reality (XR)
2. 3D—holographic imaging
3. 5D—communications
4. Visible light communications
5. Smart clothing and wearables
6. Autonomous vehicles
7. Indoor positioning systems
8. Massive M2M communications

Almost all of the next-generation of smart devices are virtualized and deployed in the cloud for seamless data transition [10, 11] to support various data-intensive applications. The majority of mobile devices will follow cloud principles like virtualization, etc. Mobile devices will also follow the agile method for efficient management and network orchestration (MANO) [12–16]

The Internet of Everything (IoE) depends on high-speed broadband access for effective communication that forms the core objective of 5G networks [17, 18]. For example, high-frequency data access is not available for applications such as VR, AR, autonomous systems, etc. These limitations are satisfied by 6G networks [19, 20]. Indeed, 6G improves the limitations faced in 5G. The acronyms used in this chapter are shown in Table 3.1. Table 3.2 shows the

Table 3.1 Acronyms Used

Acronym	Derivation
VANET	Vehicular adhoc networks
ID	Identity
RSA	Rivest-Shamir-Adleman
AES	Advanced Encryption Standard
ML	Machine learning
CL-AKA	Certificateless-based authentication and key agreement
IDS	Intrusion detection system
CEAP	Clinical-etiological-anatomical-athophysiological
UVAR	Unmanned aerial vehicle-assisted VANET routing protocol
HFMD	Historical Feedback–Based Misbehavior Detection
GDVAN	Greedy detection for VANETs
MDS	Misbehavior detection system for vehicular networks
BFD	Bidirectional Forwarding Detection
BC	Blockchain
DLT	Distributed ledger technology
CIA	Confidentiality, integrity, and availability

Table 3.2 Key Performance Indicators (KPI) of 5G and 6G

Key performance indicators	5G networks	6G
Data rate	0.1Gb/s–20Gb/s	1Gb/s–1Tb/s
Localization precision rate	10cm in 2D	1cm in 3D
Density	106/km^2	107/km^2
Error rate	<10^{-5}	<10^{-9}
Mobility rate	500km/h	1000km/h
Traffic capacity rate	10Mb/s/m^2	<10Gb/s/m^3
Latency rate	1–5ms	10–100ns

key performance indicators of 5G and 6G. Blockchain technology can assist in the applications for 6G by building trust between peer–peer entities and hence eliminating third-party intermediaries.

Blockchain is transparent, immutable, and provides tamperproof data [21]. Unpredictable conventions are associated with keeping up the trustworthiness of the disseminated blocks. 6G networks try to solve security issues to improve the potential for complex and rapid attacks [22].

This chapter is organized into various sections. Section 3.2 discusses general challenges in 6G. Section 3.3 discusses blockchain integration with 6G. Section 3.4 explores 6G network challenges, Section 3.5 states trends and goals for 6G in Section 3.6, and Section 3.7 discusses blockchain and applications in 6G networks. Finally, Section 3.8 highlights research directions for blockchain in 6G networks, followed by case study and conclusion.

3.2 GENERAL ISSUES IN 6G

Some of the general challenges or issues in 6G networks are shown in Table 3.3[23].

3.3 BLOCKCHAIN TECHNOLOGY INTEGRATION WITH 6G

Blockchain technology is one of the first innovations in technology to solve the issues related to 6G networks (Section 2). High-speed Internet connectivity says 5G or 6G will help improve the quality of service (QoS) and quality of experience (QoE) for telesurgery, Healthcare 4.0, Industry 4.0, virtual reality augmented reality, and smart education, among other applications. Figure 3.1 shows the interaction of 6G network with blockchain and how any network based applications can take benefit of adopting blockchain.

3.3.1 Intelligent Resource Management

With the exponential expansion of data in the future, resource management and decentralized computation require compatible infrastructure, which is a

Table 3.3 General Issues in 6G

Category	Challenges	Description
Connectivity in future systems	Scalability	With IoT, information is produced at monstrous rates. In this manner, it is hard for 6G frameworks to handlesuch uncommon traffic requests.
	Communication with the least inactivity	Ongoing communication in healthcare systems needs to have strong exactness with close to zero postponements.
	Higher throughput	Connecting billions of devices beyond 5G requires base stations to handle huge volumes of data in real time.
	Synchronization	Networks require protection and security from communication and require synchronization.
Security in future computing ecosystems	Confidentiality	Future computing requires lightweight encryption algorithms for low-weight IoT devices.
	Integrity	Future systems should provide security from the massive volumes of data from unauthorized users during transit.
	Availability	Service availability should be able to secure for systems beyond 5G.
	Authentication and access control	Access control mechanisms should provide provision for data access for various users.
	Audit	Auditing for enhancing 5G ecosystem security poses a significant challenge.
Higher data consumption		5G systems require higher data rates for applications such as VR, AR, etc.
Restricting resources		The computational, limit, and cryptographic impediments are troublesome with standard reception, as there might be odds of deviation.

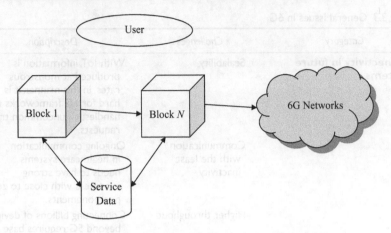

Figure 3.1 Interaction of 6G network with blockchain service data.

significant part of 6G. Zhang et al. [24] have proposed Edge intelligence and IoT framework beyond 5G. Similarly, the intelligent network architecture is built using blockchain technology where transactions are done using smart contracts [25] created using the game theory based on spectrum sharing algorithms. The use-cases for blockchain are for deep reinforcement learning, efficient resource management, and energy management [26].

3.3.2 Security Enhancement Features

1. *Privacy*: Privacy information is important for huge volumes of data generated in 5G networks.
2. A security preservation scheme was proposed by Fan et al. [27] for 5G networks.
3. *Authentication and access control*: Yang et al. [28] introduced an access control mechanism in the cloud radio for 5G networks.
4. *Integrity*: Adat et al. [29] introduced blockchain for preventing pollution attacks.
5. Ensuring the integrity of information in 5G vehicular networks over blockchain was proposed by Ortega et al. [30].
6. *Availability*: Rodrigues et al. [31] introduced a DDoS replication system to prevent DDoS attacks. A combination of blockchain technology and SDN was implemented to ensure security from distributed denial of service (DDoS) attacks, data protection, and access control [32].
7. *Accountability*: The framework's security, surveillance, and organization can be executed through blockchain-based 5G networks.

3.3.3 Versatility

Flexibility is a huge essential in 5G and future biological systems. The limitations in scalability can be eliminated in the blockchain and smart contracts. Edge computing and fog computing will improve the administration qualities and, later on, organize biological systems. 5G networks connect many users from rate-centric enhanced mobile broadband (eMBB) administrations to URLLC [33] and mobile-to-mobile communications [34, 35]. Security is a key concern in 5G networks. 6G, with more improved services, will try to solve security concerns and help in building a trust model into the networks [36].

3.4 6G NETWORK CHALLENGES

In the present world scenario, emerging technologies are growing very fast to fully automate services. The rapid growth in various outstanding technologies, like artificial intelligence, cloud computing, Internet of Things, etc., leads to the increment of data and massive traffic that individuals deal with. By 2021, it is normal that 5G will be completely conveyed around the world. 5G systems won't have the ability to bring a mechanized system that provides everything as assistance and vivid experience. Although the 5G correspondence frameworks that will be discharged very before long will offer critical enhancements over the current frameworks, they won't be capable of satisfying the requests of future rising astute and mechanization frameworks. Innovative applications drive each new smart age. 6G is no unique case: It will result from an unparalleled advancement of empowering new applications and creative examples that will shape its introduction targets, while, on a very basic level, renaming standard 5G organizations. The most significant prerequisite for 6G remote systems is the capacity to deal with gigantic volumes of information and extremely high information rate availability per gadget. Investigation practices on 6G are in their basic stages. Various researchers have described 6G as B5G or 5G-plus. Figure 3.2 shows the evolution of network technologies from 1G to 6G. In the present digital era, vast amounts of data are generated from web applications, smartphones, etc. The eruptive and versatile web innovation is the impetus empowering and engendering different best-in-class client-characterized administrations, for example, online shopping, smart homes, and online gaming. 6G networks should provide high-speed data rates to the customer.

This is possible through the successful utilization of the range in the THz framework. Applications like 3D gaming and the like require high-speed Internet access, which is possible only through 5G or 6G. For providing reliability, blockchain technology shows the way. Counterfeit information and AI will expect a meaningful activity in association and structure-level courses of

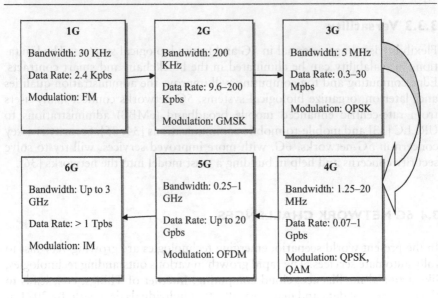

Figure 3.2 Evolution of network technologies (1G–6G).

action of 6G far off the frameworks. New access strategies will be required for extremely massive machine-type exchanges. Equalization and duplexing plans past quadrature amplitude modulation (QAM) and orthogonal frequency division multiplexing (OFDM) must be made available at THz frequencies.

Ensuring security at all levels is the main feature of 6G networks. The most important security may be provided in the physical layer. Within the 6G time frame, it will be possible to make data markets as needs are identified, and security protection is a key engaging impact for future organizations and applications. 6G isn't just about moving data around. It will wind up being an organization arrangement, including corresponding organizations where all unequivocal customer estimation and information may transfer to the edge cloud. The sixth-generation network is very useful for low-latency connectivity in the intensive intelligent society for complete automation technology [37].

3.4.1 Connectivity in Future Systems

The network is the capacity to associate frameworks or application programs. In a perfect world, these associations are built up without requiring numerous progressions to the applications or the frameworks they run. Application projects may need to speak with one another to finish exchanges or adjust assets at an establishment successfully.

3.4.1.1 Scalability

Scalability is the limit of a PC application or thing (gear or programming) to continue working commendably when changed in size or volume to meet a customer's need. The mechanical IoT devotees anticipate that billions of devices will be related and later worked on modern natural frameworks with the ascent of thoughts, for example, colossal machine-type communications (MMTC).

3.4.1.2 Communication with Minimum Latency

In correspondence, dormancy is an overflow of how much time it takes for an information pack to go, starting with the one given point, then going onto the accompanying point. In a perfect world, dormancy will be as close to zero as possible, considering the current circumstances. Continuous correspondence is a huge essential in future figuring situations. The use-cases, for example, self-driving cars and AR-helped social protection structures, may require an anticipated immaterial delay in correspondence engaged in enormous scope data exchange.

3.4.1.3 Higher Throughput

Throughput refers to how much data can be moved to begin with in one area, onto the following area in a given period of the time. It is used to check the introduction of hard drives and RAM, similar to Internet and framework affiliations. 5G networks try to build a strong ecosystem capable of connecting billions of devices where sharing and data transfer should happen at faster rates continuously without delay.

3.4.1.4 Synchronization

Synchronization is a prerequisite aspect of future networks. The strategic spine structures, including power conveyance structures and vehicular frameworks, must synchronize in progressing toward exact movement. Timing synchronization is the procedure by which a recipient hub decides the right moments to test the approaching sign. Bearer synchronization is the procedure by which a collector adjusts the recurrence and period of its nearby transporter oscillator to those of the received signal.

3.4.2 Security in Future Computing Ecosystems

Most basic correspondence frameworks incorporate fundamental components of data security for control information, while information encryption, particularly client traffic information, isn't regularly obligatory. The framework or the client might apply contingencies on the application, extra encryption, and respectability coding.

3.4.2.1 Confidentiality

The symmetric key encryption computations should be lightweight enough to perform the low-force IoT gadgets, and they should ensure security during computational tasks.

3.4.2.2 Integrity

The huge volume of data generated in future systems requires data to be stored securely as it is moving at high speed in wireless networks. Integrity of the data ensures that only the authorized users can access the data.

3.4.2.3 Availability

Highly available data is the main feature of future networks. With the advancements of 5G, a huge volume of data is generated from many interconnected devices, posing the risk of DDoS attacks. The quality of the security instruments for 5G and past frameworks can't be relied on to recognize hazards and break tries.

3.4.2.4 Authentication and Access Control

The data, either in transit or at rest, requires ensuring security mechanisms to prevent malicious attacks. The customary bound-together check and access control mechanisms will limit flexibility in the monstrous cutting-edge demands anticipated in 6G. The access control mechanisms should facilitate the diversification of the future tenants in the 6G networks such that they do not cause bottlenecks in the current organization.

3.4.2.5 Audit

An audit is required to evaluate the consistency of the nodes in the networks. For the anticipated raised security standards, the significant pack level audit may need to recognize and hail the lead of those nodes. The reviewing of endless occupants will be tested from the perspective of maintaining security.

3.4.3 Higher Data Consumption

The applications, for example, VR, holographic correspondences, 16K video, and 3D ultra video, require higher data rates and data usage. To perform some automation through wireless technology, communications need higher data consumption to implement the latest techniques, including artificial networks and other technologies discussed in preceding sections.

3.4.4 Resource Restrictions

Restricting resources are anticipated to limit the capacities of cryptographic estimations and eventually lead to deviation from the standard systems. The legal security mechanisms should be more persevering with such device resource requirements.

3.5 6G: TRENDS

6G trends are illustrated in the subsequent sections [38].

3.5.1 Trend 1: More Range, Greater Unwavering Quality

All the sixth-generation wireless communication applications will require a larger bit rate than that of the 5G spectrum since it has to handle more data with low latency. To deal with the large applications that run a huge amount of data, 6G must deliver the data more effectively. The requirement for higher reliability to support the strength of 6G and to meet the highest frequencies is becoming ever more challenging.

3.5.2 Trend 2: Spatial to Ground-Level Spectrum and Its Vitality Productivity

6G will assault the field and airborne clients, including workstations and both XR/BCI frameworks and moving airplanes. The 3D 6G nature requires advancement toward the meaning of volumetric instead of spatial (areal) transmission capacity. The data transfer rates vary from 2G (bps) to 3G (bps/ Hz), 4G (bps/Hz/m^2), and 5G (bps/Hz/bps/Hz).

3.5.3 Trend 3: Advancement of Insightful Surfaces and Conditions

Current cell frameworks and past base stations (of different sizes and shapes) were utilized for transport. We are experiencing a revolt in surfaces working electromagnetically (e.g., using metamaterials) and utilizing artificially developed walls, parkways, and even entire houses. For example, the Berkeley wallpaper venture is a case in point.

3.5.4 Trend 4: Tremendous Accessibility of Nano Information

Computerized progress will quickly move from concentrated, enormous information to colossal "small-scale" information spread through the short term. 6G systems mean coordinating gigantic and small databases through their applications to upgrade, organize limits, and confer inventive abilities. This example inspires new methods of AI that go beyond regular enormous information investigation.

3.5.5 Trend 5: From Self-Organizing Networks (SON) to Self-Sustaining Networks

SON was scarcely fused into 4G/5G, organized for the most part due to a deficiency of genuine necessities. CRAS and DLT advances, be they as they may, presented a quick requirement for a keen SON to oversee the arrangement of tasks, assets, and enhancements. 6G will require a change in perspective from the old-style SON, whereby the system simply adjusted its capacities to explicit condition states, to a self-continuing system (SSN) fit for keeping up with its key execution pointers (KPIs), ceaselessly, in profoundly unique and complex situations originating from the rich 6G application areas. SSNs must not exclusively have the option to modify their jobs; however, they should likewise have the option to help them use and utilize capital (e.g., vitality creation and misuse range) to support solid, long-haul self-governing KPIs. The SSN capacities must use the ongoing AI innovation unrest to make AI-fueled 6G SSNs.

3.5.6 Trend 6: Convergence of Communications, Computing, Control, Localization, and Sensing (3CLS)

The cellular systems of the previous five decades had one select capacity: remote correspondence. In any case, 6G would challenge this rule by combining various capabilities (i.e., joint and concurrent contribution) that incorporate parallel computing, following, localization, and detecting. We visualize 6G as a multi reason arrangement fit for offering various 3CLS types of assistance that are particularly alluring and even required for various applications. For example, force, position, and calculation are inalienable in XR (extended reality), CRAS (Centralized Remedial Action Scheme), and DLT (distributed ledger technology). Also, detecting administrations will permit 6G frameworks to convey a 3D plan of the radio condition across various frequencies for clients. Additionally, 6G is organized to execute and deal with 3CLS administrations in an exacting manner. Note that developments identified with past patterns will slowly be made to work for 6G frameworks to make 3CLS prompt.

3.5.7 End of Personal Digital Assistant (PDA) Period

Existing gadgets are not feasible for upgrading to 4G and 5G technologies. The latest devices capable of adapting to smart technologies will play a vital role in the coming generation. Smartwatches and the like will take on the aspect of 6G by pushing off the smart era.

3.6 BLOCKCHAIN AND 6G NETWORKS

Blockchain is a means of recording information with the end goal of making it inconvenient or hard to change, hack, or cheat the system. A blockchain is fundamentally a propelled record of replicated trades scattered over the entire arrangement of PC systems on the blockchain. Each square in the chain contains different professions, and each time another transaction occurs in the blockchain, a record of that trade is added to each member's catalog. The decentralized database managed by various individuals is known as distributed ledger technology (DLT). Blockchain technology executes a dispersed record: a protected, decentralized type of database where no single gathering controls. It offers a tough, solid, straightforward, and decentralized method of recording and controlling information over all the hubs of a system of investing individuals that need to stay up with the latest. Blockchain is most popular as the premise of Bitcoin, the privately advanced "cryptographic money" that can work as cash, regardless of not being issued by any legislature. Notwithstanding, disseminated records have a lot more employments. Currently, different administrations are working on the Bitcoin blockchain—free blockchains with their digital currencies (for example, the Ethereum and Ripple) and circulated records with no local cash [39].

Blockchain is a DLT in which trades are recorded with a constant cryptographic imprint called a hash. Figure 3.3 shows the comparison between DLT and blockchain-crypto digest. Each transaction creates a hash. Each square suggests the previous count, and the two together make the blockchain. A blockchain is practical because it is spread over various PCs, all of which have copies of the blockchain. Shared fiscal and natural variables, given headway in the square chain, will quicken the advancement of decentralized frameworks. Through data encryption, timestamps, keen understanding, and other specific strategies, the common (P2P) trades can be cultivated. The presentation of blockchain innovation can lessen the hindrances to section-exchanging frameworks while guaranteeing straightforwardness and speedy access to resource exchanges. The blockchain is a rising development reliant on decentralized preparing and data stockpiling, ensured by a blend of cryptographic stamps and appropriated sharing instruments.

Blockchain is a key formative innovation that can show the genuine capacity of sixth-era remote system frameworks. For the planned mechanization

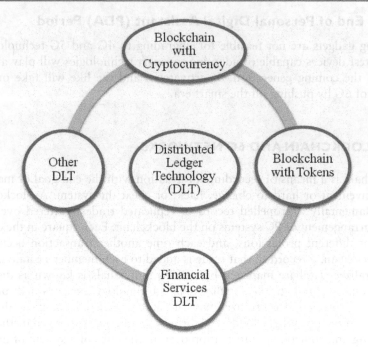

Figure 3.3. DLT vs. blockchain–crypto digest.

condition, the system's saving guideline is troublesome given the anticipated huge connectedness prerequisite. Blockchain is one of the most recognizable advances in discharging the ability of 6G's network functionalities. Despite its limitations, blockchain can solve the challenges of 6G. Blockchain can be the data backbone of an appropriate asset framework by arranging clients and makers in an open, straightforward market and superseding the informal- town boundaries to broadcast the assets and quicken the pace of exchanges. Blockchain has extended the horizon of resource exchanging of fixed resources, for example, authorized spec-tram and registering equipment. Blockchain opens up straightforward and circulated data reorganization, which can profit all aspects of businesses, obliging a scope of centralization utilizing various CMs. Blockchain has enabled straight and disseminated data reconstruction, which can benefit all ventures, obliging all contents of centralization using multiple CMs.

With regard to using blockchain development in 6G, the enormous organization of the blockchain may initiate important developmental advances for related businesses and all other segments of the economy. In this exceptional circumstance, blockchain is a remarkable tribute to both IoT and D2D correspondences since it furnishes a straightforward system with expanded interoperability, dependability, effectiveness, and adaptabilities. For instance, blockchain might be utilized to direct range sharing concerning asset rulers and to archive all solicitations for extended use and rent. Also, it can supply the motivation for gadgets to share and exchange assets, inasmuch as current conventions do not have the impetus to do so. Organizing the blockchain into IoT and D2D correspondences can lead to the arrangement of more agreeable and confidential conditions.

Moreover, this structure can be applied similarly to range sharing, where a client leases range to another, making a powerful system condition and improving the sufficiency of range. Spectrum management permits multiple categories of clients to securely have a similar recurrence band, making increasingly productive utilization of the radio range accessible for communication. In range sharing, an essential client negotiates leases with other clients to utilize some-degree of its range. Blockchain innovation can improve the sharing of security by forestalling the altering of the lease records. 6G network communication works with fast connectivity to wireless devices. Due to the connectivity of gadgets to gadgets, high volumes of information might be stored on client devices. Such storage may reduce traffic and improve network speed and reliability. Since the communication content may contain sensitive client information, thereby surfacing privacy issues, cache requesters need a trustworthy cache provider. Blockchain can build trust among cache requesters and providers since it maintains the data in a secure mode. Blockchain can provide a common open-source wireless intelligent network architecture used in the market to exchange the resources of users, spectrum owners, infrastructure owners, and Internet service providers. The system asset, the board, is becoming an ever more testing issue in telecommunications.

3.7 APPLICATIONS OF BLOCKCHAIN IN 6G NETWORKS

Applications of blockchain in 6G networks are related to Industry 4.0. The important aspects of blockchain in 6G networks are discussed in the following sections [40].

3.7.1 Modern Applications beyond Industry 4.0

In 6G, the cutting-edge applications will be noteworthy drivers for mishandling 6G capacities. The key characteristics of blockchain and these difficulties are particularly pertinent to current conditions. For instance, holographic interchanges for modern use-cases, such as remote support or the gigantic availability of mechanical production hardware, require decentralized structures that are simultaneously reliable. Blockchain can define these limits when they are consolidated into these applications or use-cases. Notwithstanding, there are also significant exploration challenges regarding blockchain-based arrangements, specifically inactivity and versatility. They are impressive due to the rigid execution prerequisites in modern applications that are more substantial for mechanical systems and IoT.

3.7.2 Consistent Environmental Monitoring and Protection

With its consistent environmental monitoring and protection, blockchain provides decentralized support, normally identifying applications that on an overall scale with 6G. Such capacities can serve use-cases, such as clearly defined urban regions or transportation, similar to biological protection for a green economy. Blockchain, in like manner, empowers secure data sharing among parties (reaching out from IoT devices to affiliations). Such monstrous scope for detecting and information sharing arrangements entrusted to blockchain are vital for natural checking. Additionally, combined and shared learning, actualized through blockchain, bolsters the information, examination, and derivation forms of biological insurance in a decentralized way.

3.7.3 Smart Healthcare

Smart healthcare comprises numerous members, such as experts and patients, clinical centers, and assessment foundations. It characteristically incorporates various estimations, including sickness neutralization and checking, finding, and treatment; a crisis facility on the board, a prosperity dynamic, and clinical investigation. Information developments, such as IoT, convenient Internet, disseminated figuring, enormous data, 6G, microelectronics, and artificial thinking, alongside present-day biotechnology, establish rigorous human administrations. These developments are commonly employed in all aspects of clearly defined social protection. From patients' viewpoint, they will utilize portable gadgets to track their success accurately and check for therapeutic assistance from associates, even as homes conduct affairs with far-off organizations. From the professionals' perspective, there is a structure of watchful professional decision making, using solid procedures to assist and even enhance confidence in the decision making. Professionals can administer clinical information through an organized information vehicle that consolidates the laboratory information management system, such as Picture Archiving and Communication Systems (PACS), electronic medical records, etc. Continuously precise clinical techniques can be cultivated by means of cautious robots and the development of mixed reality. Securing 6G healthcare should go one step further than 5G frameworks to unwind the in-patient problem [41].

The more significant and omnipresent coordination of square chains in future frameworks can propel current structures of restorative administration and improve execution for better decentralization, security, and insurance. Among these issues, protection is paramount. What's more, with regard to human administration decency, data can be had due to the capacity of the perpetual square chains. Specifically, square chains can be empowered with customer-controlled insurance and secure data gathering even without a common trust in the outcome. GDPR (general data protection regulation) commands are huge drivers in Europe that will eventually become progressively stiffer in the years ahead. Better decentralization will involve greater security, particularly as far as openness toward the fundamental specification of protection is concerned.

3.7.4 Decentralized and Trustworthy 6G Communications Infrastructure and Solutions

There are plenty of open doors for misusing blockchains in 6G foundations for execution gains or empowering new administrations/use-cases. To be specific:

1. *Decentralized system in the organizational structures*: In the decentralized blockchain-based system, the board will give the better asset to the executives and progressively productive framework.
2. *Valuing, charging, and charging system administrations*: Blockchain can empower charging and charging without a combined framework that is a progressively adaptable and productive design, in contrast with customary frameworks.
3. *Authentication, authorization, and accounting (AAA)*: At the point, to a great degree, accessibility in heterogeneous and isolated parts in the framework is set up in the 6G framework, and AAA limits ought to be decentralized and extensively more rigorous about organization movement. For example, key association and access control instruments can be offloaded to blockchain stages for greater adaptability and straightforwardness.
4. *Service-level agreement (SLA)*: 6G frameworks will develop treated and precut framework designs like 5G networks, but they will be huge in scope. In addition, these systems are expected to serve an exceptionally wide range of use-cases with different assistance-level certifications. Along these lines, the SLA is a significant framework prerequisite. Blockchain secures the SLAs and thereby ensuring security of data.
5. *Spectrum sharing*: Limited improvement and rangeability for 6G radio access aren't evident with united organizational structures and abnormal sharing plans. Blockchain and clear understandings can decrease the range of sharing related to coordinated efforts and straightforwardness issues.
6. *"Extreme edge"*: 6G structures need to help the spatial interpretation of many center associations from the cloud to the edge systems for

accomplishing incomprehensibly low-latency correspondences and dual systems. Reliable coordination and direct asset accounting can be practiced with square chains in these frameworks.

3.8 RESEARCH DIRECTIONS

There are several opportunities to explore in 6G and blockchain technology. 6G and blockchain, once consolidated, offer numerous possible focal points, and one of them is providing endogenous security. The plan of 6G's security framework will receive a conveyed and decentralized structure in order to adjust it to the current remote system. Having the comparative attributes of minimization and shared design as in 6G, blockchain innovation is a basic device in improving 6G's endogenous security execution. The examination capability of 6G is enormous, with differing software engineering and media transmission research roads. The most open doors to unmistakable exploration for 6G with blockchain innovation are examined in this segment.

3.8.1 Internet of Everything (IOE)

The Internet of Everything (IoE) is a specific concept that alludes to Web-related devices and consumer products and is packed with modern, expanded highlights. This is a way of thought with regard to potential developments in a wide variety of computers, apps, and worldwide Internet-related issues. The invigorated advancement of 6G administrations and utilization of blockchain to store IoE data would require another layer of security that developers would need to create in inducting participants to the framework. Blockchain gives a considerably greater level of encryption that makes it, in every way that matters, hard to overwrite existing data records. The circulated record innovation (DLT) in a blockchain system is painstakingly structured, and this avoids the prerequisite for trust among the included social transactions. Blockchain can empower the quick handling of exchanges and coordination among billions of associated gadgets. As the number of interconnected devices increases, the scattered records approach offers an appropriate response to help treat the incalculable number of trades. In general, IoE will call for a reconsideration of business strategies and plans of activity.

To start with, the relevant forms need to be optimized and automated for this computerized innovation. Second, to utilize the automation, new business plans of action need to be conceived. Exploring a market, it would be interesting to see the aftereffects of the different possibilities when addressing IoE. There would be a real need to compete at an extraordinary pace and with uncommon agility within the company. At the same time, the effect of using blockchain-based innovations for inspiration would drive interoperability between different organizations [42].

3.8.2 Smart Contracts in Ethereum

Ethereum is the world's second-largest cryptocurrency platform created with the help of blockchain technology. With Ethereum, decentralized applications can be made, which are called smart contracts and which are immutable. Smart contracts can be created and deployed with confidence. Since the decentralized applications are immutable, they have great security. Data in the Ethereum network is stored under high security, so tampering with it may not possible for hackers [43].

3.8.3 Data Storage and Analytics

By completing the IoE, a colossal number of things and articles will reliably drive nonstop surges of new information. Subsequently, regardless of satisfactorily and effectively gathered and decentralized data, headways are required for the accumulations. Blockchain-empowered advances can play an important job here. Integrating the 6G wireless communication services with blockchain technology will help in data storage, which is highly secure with its hashing and DLT technology. Extensive research is still underway in this sector for integration into emerging technologies like cloud and edge computing. Research on strategies for information examination will be urgently required to break down this huge load of information and extract the basic components for productive and precise utilization. The four fundamental classifications of strategies, for the most part, rely on the nature of use: elucidating examination, demonstrative investigation, prescient examination, and prescriptive investigation. Once more, it will be intriguing to examine the conceivable outcomes from consolidating these information examination techniques with disseminated blockchain-based information stockpiling. The upside of smart agreements is that they can be abused to mechanize the procedures.

3.8.4 Artificial Intelligence and Machine Learning

Artificial intelligence (AI) is a wide-ranging type of programming, building machines that are fit to perform tasks that otherwise consistently require human intervention. AI is an interdisciplinary science with a range of methods. AI trends and major research bring out a change in perspective in any direction that applies to the software group. In the early stages of communication, artificial intelligence was not implemented so much, whereas in 5G network communications, the integration of artificial intelligence occurred but to limited use-cases. In the 6G network, automation can be more fully implemented. Through wireless technologies and artificial intelligence, automation becomes part of more emerging fields. Blockchain aims to strengthen the systems developed with the aid of artificial logic, and AI approaches have tended to enhance channel labeling, broadening some form of acknowledgment. Assessment is already an important process in both of these areas and needs more evaluation. At the system layer, the currently applied 5G advances, like SDN, NFV, and slices, should also be improved to ensure powerful, flexible, and self-learning adaptive designs ready to support the increasingly bewildering and heterogeneous systems, in order to keep them reliably and strongly progressing. Artificial intelligence will take on an urgent job in 6G correspondence. For instance, AI calculations could be utilized to assign base station assets and accomplish near-ideal execution productively. Smart materials positioned on surfaces, such as roadside structures or roadway lights, could be utilized to detect remote conditions and to alter changes in radio waves.

Furthermore, advanced learning methods could be utilized to improve the exactness of indoor situating. In this process, blockchain will help to secure the decision process of the algorithms and models developed with the help of machine learning. The research challenges and issues in ML and BC are shown in Figure 3.4 [44].

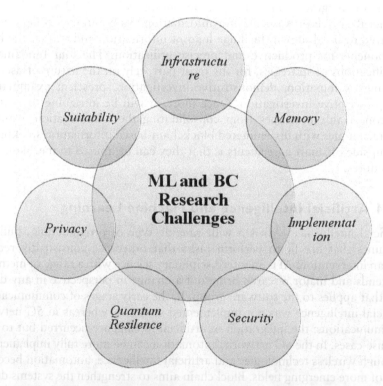

Figure 3.4 ML adoption in BC—research challenges and open issues.

3.9 CASE STUDY: LOGISTICS SUPPLY CHAIN

The supply chain is involved in the intricate processes of the creation and distribution of products. Contingent upon the type of good, the supply chain can incorporate numerous stages, geographic areas, ledgers and installments, people, substances, and methods of transport. In this way, the availability of provisions can be extended. Because of the unpredictability and the absence of straightforwardness of conventional chains, it is up to the incredible enthusiasm of the partners engaged in the coordination cycle to become acquainted with and create blockchain innovation and to upgrade coordination measures in the flexible chain, making them more practical. Tracking products through blockchain can improve the dynamic process, with the end product being an all the more fulfilling service for the end client. Blockchain innovation has the potential for making new logistics services, as well as new plans of action. As a general innovation, blockchain is intended to accomplish decentralization, constantly shared activity, transparency, irreversibility, and uprightness in a broadly relevant way. It tends to be utilized for any information trade, regardless of whether it is contracted, connected with shipments, or monetary transactions. Each activity is caught in the block, and the information is circulated over numerous nodes, making the framework straightforward. Each block interfaces with the one prior and then afterward, which makes the framework more secure. Blockchain can build the productivity and straightforwardness of the supply chain and dramatically influence every logistic process, from capacity to conveyance and installation.

One impediment is its performance. The confirmation of each exchange requires the affirmation of each node in the network, which takes considerably more time than in the centralized framework. The time factor is, of course, one of the important things that should be considered while implementing any type of application. 6G networks will help blockchain technology offset the time factor lacking in the centralized system.

As shown in Figure 3.5, logistics organizations have followed the traditional process for the past few years. Here, from the business perspective, logistics is used to ship products from one origin point (the source) to consumption point (the destination). Many logistics and supply chain businesses are concluding that it is too hard to maintain physical records. Yet, in terms of customer

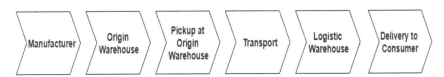

Figure 3.5 General process of logistics.

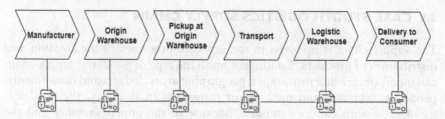

Figure 3.6 Blockchain process in logistics,

satisfaction and business growth, winning customer trust is challenging for organizations.

Rapidly changing customer desires and the cost factor make the traditional supply chain process complex and redundant. Figure 3.6 shows the process of implementation of the blockchain in logistics. An ideal way that blockchain can help logistics is to comprehend its failures. Simultaneously, pretty much every component of a supply chain has a huge number of choices to manage. Intermediaries, transporters, and others focus on effectiveness instead of stalling out while picking the ideal choice. At the head of all that, huge amounts of documentation need to be finished. The documentation process leads to more complex situations in end-to-end transportation [45].

Smart contracts are used to implement blockchain logistics to store each end-to-end transportation step from source to destination, i.e., manufacturer to the consumer. The proposed smart contract includes a progression of functions utilizing the hashing technique. Each smart contract, once sent, has its blockchain address, so a client can call on the smart contract by starting an exchange and passing the function hash code into the agreement.

3.9.1 Process for Product Request: Request_Product ()

This function rises when the consumer orders a product from the manufacturer. The function carries out the order specifications like product_id, location_address, user_id, ordered_date into a smart contract by using LaTeX.

Request_Product (product_id, location_address, user_id, ordered_date)

Once the consumer orders a product, the request_product () raises the request by sending all the required information to the manufacturing unit, stored in blocks. Whenever the manufacturing team has a bid, the self-executable blockchain algorithm is used to estimate the travel time if the product is deliverable. Otherwise, the order request is declined.

3.9.2 Process for Tracking Goods: Product_location ()

The product_location () function is used to track the product with the help of the parameters given by the user. W the input parameters, this function retrieves the exact non-tampered-with information from a smart contract.

Product_location (order_id)

3.9.3 Process for Raising Request: Support ()

The support () function is used to raise any type of request from the user to the support team of the supply chain. This function updates the proposals given by the user. For instance, the user who wants to change the details of the delivery location has to call the support function with the respective required information. If user is satisfied with the service, the changed details are updated into the smart contract using LaTeX. Figure 3.7 shows the complete flow of the request access.

Support (order_id, request_details)

Whenever the manufacturer gets the order from the consumer, the order is shipped from the manufacturing unit to the consumer location. The entire

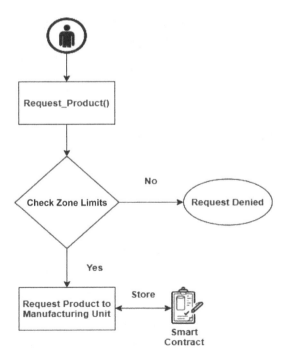

Figure 3.7 Flowchart representing the request access.

process is stored in the smart contracts using blockchain. So smart contracts have all the traveling details of the particular product from origin to the consumer as a proof-of-work document in LaTeX format. The consumer who wants to check the exact location of the ordered product calls the product_ location () function with the required details as parameters.

After receiving the respective product details, the function checks whether the given product details are matched with the user details. If the parties are matched, then function product_ location () retrieves the data of the current location of the respective product; otherwise, the request is denied. Similarly, if the consumer raises any request, the system has to call the support () function. Then the function verifies the details to complete the request; otherwise, the request is denied. Smart contracts maintain the whole process in a document that is tamper free. The time factor is the main important factor since no customer wants to wait very long to complete the operation. So the 6G wireless network is used to improve time efficiency and to maintain the records. Data retrieval gets easier when the network is more efficient. So the supply chain is more efficient with blockchain-enabled sixth-generation wireless networks.

3.9.4 Process for Update Location: Update ()

The update () function is used to update the traveling details of a particular product for the respective smart contract in the form of LaTeX by encoding the order details. All the transport information from the manufacturing unit (source) to consumer (destination) is updated into a smart contract by calling the support function with location and order_id parameters. For every transformation, the smart contract is updated recursively.

Update (order_id, Location_details)

6G-enabled blockchain in the supply chain process may reduce the hard copies (paper-based documents) through automation. 6G technologies help improve the most challenging tasks for blockchain in logistics like time efficiency and data transfer. Since blockchain is a most secure technology, this system can improve both manufacturer and consumer trust. It offers transparency, immutability, and security. Since the blockchain is a decentralized application, no one can tamper with the data in the smart contracts. For these reasons, it is an immutable and more secure process. See Figure 3.8.

3.10 EXPERIMENTAL SETUP

The proposed system is implemented on the Remix, executed, and assessed on the Ethereum Main net. The smart contract address in which the proposed work

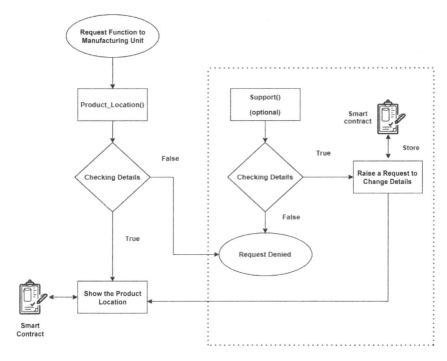

Figure 3.8 Flowchart representing the process.

Table 3.4 Gas Consumed for Smart Contract Deployment

	Gas limit	Gas price (Gwei)
Deployment of smart contract	2,368,748	2.5

is accomplished is 0x148e6e91AfE1DbCD7Ac3C5DE1881E2435a84BC54. All exchanges executed utilizing the proposed smart contract are recorded in this address and are freely accessible on Etherscan. Etherscan permits anybody to research the Ethereum blockchain for exchanges, addresses, and various occurring exercises. Application binary interface (ABI) is used to better interact between the deployed smart contract and the user. ABI can build the interaction platform between the user and smart contract for executing the required functions only if the user has the required information about the smart contract address. The gas consumed to deploy the smart contract using Ethereum is shown in Table 3.4. The gas required to deploy the functions in smart contracts using Ethereum is shown in Table 3.5.

In the Ethereum organization, gas is a unit of cost for executing a particular exchange. The gas limit is the greatest measure of gas a client is willing to

Table 3.5 Gas Limit for Smart Contract Deployment

Function	Gas limit	Gas price (Gwei)
Request_product()	259,867	8
Product_location()	358,969	15
Support()	315,749	17
Update()	485,968	22

Table 3.6 Applications of Blockchain in the Supply Chain across Various Sectors

	Scalability	Privacy	Interoperability	Auditing	Product provenance	Latency	Visibility	Disintermediation
Finance	Yes	Yes	Yes	Yes	Yes	Yes		Yes
Healthcare	Yes			Yes				Yes
Manufacturing	Yes			Yes	Yes		Yes	
IoT	Yes	Yes	Yes	Yes		Yes	Yes	
Social service		Yes	Yes	Yes			Yes	
Shipping		Yes	Yes	Yes	Yes		Yes	Yes
Food and agriculture	Yes			Yes	Yes		Yes	Yes
Education			Yes	Yes				Yes
E-commerce	Yes			Yes	Yes		Yes	Yes

spend on an exchange, and the gas cost is the Gwei cost per unit of gas. For every organization, or capacity call, Ethereum proposes a specific gas limit required for the exchange, which depends on the smart contract necessities and may very well be adjusted. On the off chance that a lower gas limit is utilized, the agreement arrangement, or the capacity call, will be dropped, so using it as far as possible or even incrementing it is encouraged.

3.10.1 Blockchain Applications in the Supply Chain across Various Sectors

Blockchain plays a key role in the supply chain across various sectors. Table 3.6 shows the applications of blockchain in the supply chain sectors [46].

3.11 CONCLUSION

This chapter discussed the improvement of the performance of blockchain technology through adopting high-performance 6G network communications. The digital transmission of blockchain-based applications drastically improves the performance in terms of speed, quality of service, and reduced network latency. The chapter also discussed the adoption of 6G under various real-time applications to improve performance and enhanced security. The seamless network bandwidth and the unique performance metrics related to the 6G network drastically complete the given task on time without delay. The technology is utilized under a specific domain like a network-related problem. It also covers a broad range of applications, including machine learning and virtual reality.

REFERENCES

1. B. Aazhang et al, "Key drivers and research challenges for 6G ubiquitous wireless intelligence," (white paper), 09 2019.
2. M. Z. Chowdhury et al, "6g wireless communication systems: Applications, requirements, technologies, challenges, and research directions," arXiv preprint arXiv: 1909.11315, 2019.
3. I. Union, "IMT traffic estimates for the years 2020 to 2030," Report ITU-R M. 2370–0, ITU-R Radio communication Sector of ITU, 2015.
4. E. C. Strinati et al, "6g: The next frontier," arXiv preprint arXiv: 1901.03239, 2019.
5. M. Piran et al, "Learning-driven wireless communications, towards 6g," arXiv preprint arXiv: 1908.07335, 2019.
6. F. Tariq, M. Khandaker, K.-K. Wong, M. Imran, M. Bennis, and M. Debbah, "A speculative study on 6g," arXiv preprint arXiv: 1902.06700, 2019.
7. J. Fleetwood, "Public health, ethics, and autonomous vehicles," *American Journal of Public Health*, vol. 107, no. 4, pp. 532–537, 2017.
8. S. Dang, O. Amin, B. Shihada, and M.-S. Alouini, "From a human-centric perspective: What might 6g be?" arXiv preprint arXiv: 1906.00741, 2019.
9. W. Saad, M. Bennis, and M. Chen, "A vision of 6g wireless systems: Applications, trends, technologies, and open research problems," arXiv preprint arXiv:1902.10265, 2019.
10. Z. Zhang, Y. Xiao et al, "6g wireless networks: Vision, requirements, architecture, and key technologies," *IEEE Vehicular Technology Magazine*, vol. 14, no. 3, pp. 28–41, 2019.
11. N. H. Mahmood et al, "Six key enablers for machine type communication in 6g," arXiv preprint arXiv: 1903.05406, 2019.
12. S. Yrjola, *Decentralized 6g Business Models.*6G Wireless Summit,2019.
13. X. Li, M. Samaka, H. A. Chan, D. Bhamare, L. Gupta, C. Guo, and R. Jain, "Network slicing for 5g: Challenges and opportunities," *IEEE Internet Computing*, vol. 21, no. 5, pp. 20–27, 2017.
14. D. C. Nguyen, P. N. Pathirana, M. Ding, and A. Seneviratne, "Blockchain for 5g and beyond networks: A state of the art survey," arXiv preprint arXiv:1912.05062, 2019.
15. A. Nag, A. Kalla, and M. Liyanage, "Blockchain-over-optical networks: A trusted virtual network function (VNF) management proposition for 5Goptical networks," in *Asia Communications and Photonics Conference*. Optical Society of America, 2019, pp. M4A–222.

16. T. Hewa et al, "The role of blockchain in 6G: Challenges, opportunities, and research directions," *Research Gate*. www.researchgate.net/publication/338831183.

17. I. Ahmad, T. Kumar, M. Liyanage, J. Okwuibe, M. Ylianttila, and A. Gurtov, "Overview of 5g security challenges and solutions," *IEEE Communications Standards Magazine*, vol. 2, no. 1, pp. 36–43, 2018.

18. W. Saad, M. Bennis, and M. Chen, "A vision of 6g wireless systems: Applications, trends, technologies, and open research problems," arXiv preprint arXiv: 1902.10265, 2019.

19. L. Lov'en, T. Lepp̈anen, E. Peltonen, J. Partala, E. Harjula, P. Porambage, M. Ylianttila, and J. Riekki, "Edgeai: A vision for distributed, edge native artificial intelligence in future 6G networks," *The 1st 6G Wireless Summit* (Levi, Finland), pp. 1–2, 2019.

20. B. Aazhang, P. Ahokangas, L. Lov'en, et al, "Key drivers and research challenges for 6G ubiquitous wireless intelligence (white paper)," Oulu, Finland: 6G Flagship, University of Oulu, 1 ed., 2019.

21. Y.-C. Liang, "Blockchain for dynamic spectrum management," in *Dynamic Spectrum Management*. Springer, pp. 121–146, 2020.

22. T. Nguyen et al, "Privacy-aware blockchain innovation for 6G: Challenges and opportunities," *Research Gate*. www.researchgate.net/publication/339271615.

23. A. Biral, M. Centenaro, A. Zanella, L. Vangelista, and M. Zorzi, "The challenges of m2m massive access in wireless cellular networks," *Digital Communications and Networks*, vol. 1, no. 1, pp. 1–19, 2015.

24. K. Zhang, Y. Zhu, S. Maharjan, and Y. Zhang, "Edge intelligence and blockchain empowered 5g beyond for the industrial internet of things," *IEEE Network*, vol. 33, no. 5, pp. 12–19, 2019.

25. T. Maksymyuk, J. Gazda, L. Han, and M. Jo, "Blockchain-based intelligent network management for 5g and beyond," in *2019 3rd International Conference on Advanced Information and Communications Technologies (AICT)*. IEEE, 2019, pp. 36–39.

26. Y. Dai, D. Xu, S. Maharjan, Z. Chen, Q. He, and Y. Zhang, "Blockchain and deep reinforcement learning empowered intelligent 5g beyond," *IEEE Network*, vol. 33, no. 3, pp. 10–17, 2019.

27. B. Mafakheri, T. Subramanya, L. Goratti, and R. Riggio, "Blockchain based infrastructure sharing in 5g small cell networks," in *2018 14th International Conference on Network and Service Management (CNSM)*. IEEE, 2018, pp. 313–317.

28. K. Fan, Y. Ren, Y. Wang, H. Li, and Y. Yang, "Blockchain-based efficient privacy preserving and data sharing scheme of content-centric network in 5g," *IET Communications*, vol. 12, no. 5, pp. 527–532, 2017.

29. H. Yang, H. Zheng, J. Zhang, Y. Wu, Y. Lee, and Y. Ji, "Blockchain based trusted authentication in cloud radio over fiber network for 5g," in *2017 16th International Conference on Optical Communications and Networks (ICOCN)*. IEEE, 2017, pp. 1–3.

30. V. Adat, I. Politis, C. Tselios, and S. Kotsopoulos, "Blockchain enhanced secret small cells for the 5g environment," in *2019 IEEE 24th International Workshop on Computer Aided Modeling and Design of Communication Links and Networks (CAMAD)*. IEEE, 2019, pp. 1–6.

31. V. Ortega, F. Bouchmal, and J. F. Monserrat, "Trusted 5g vehicular networks: Blockchains and content-centric networking," *IEEE Vehicular Technology Magazine*, vol. 13, no. 2, pp. 121–127, 2018.
32. B. Rodrigues, T. Bocek, A. Lareida, D. Hausheer, S. Rafati, and B. Stiller, "A blockchain-based architecture for collaborative DDOS mitigation with smart contracts," in *IFIP International Conference on Autonomous Infrastructure, Management and Security*. Springer, 2017, pp. 16–29.
33. P. K. Sharma, S. Singh, Y. S. Jeong, and J. H. Park, "Distblocknet: A distributed blockchains-based secure SDN architecture for IoT networks," *IEEE Communications Magazine*, vol. 55, no. 9, pp. 78–85, 2017.
34. W. Saad, M. Bennis, and M. Chen, "A vision of 6g wireless systems: Applications, trends, technologies, and open research problems," arXiv preprint arXiv:1902.10265, 2019.
35. Y. Dai, D. Xu, S. Maharjan, Z. Chen, Q. He, and Y. Zhang, "Blockchain and deep reinforcement learning empowered intelligent 5g beyond," *IEEE Network*, vol. 33, no. 3, pp. 10–17, 2019.
36. I. Ahmad, T. Kumar, M. Liyanage, J. Okwuibe, M. Ylianttila, and A. Gurtov, "Overview of 5g security challenges and solutions," *IEEE Communications Standards Magazine*, vol. 2, no. 1, pp. 36–43, 2018.
37. B. Aazhang, P. Ahokangas, L. Loven et al, "Key drivers and research challenges for 6G ubiquitous wireless intelligence (white paper),"Oulu, Finland: 6G Flagship, University of Oulu, 1 ed., 2019.
38. W. Saad, M. Bennis, and M. Chen, "A vision of 6g wireless systems: Applications, trends, technologies, and open research problems," arXiv preprint arXiv:1902.10265, 2019.
39. K. Werbach, *The Blockchain and the New Architecture of Trust*. MIT Press, 2018.
40. W. Saad, M. Bennis, and M. Chen, "A vision of 6g wireless systems: Applications, trends, technologies, and open research problems," arXiv preprint arXiv:1902.10265, 2019.
41. S. Tian et al, "Smart healthcare: Making medical care more intelligent," *Global Health Journal*, no. 62–25, 2019.
42. L. T. Yang et al, "Internet of Everything," *Mobile Information Systems*, vol. 8, 2017.
43. A. Pinna et al, *A Massive Analysis of Ethereum Smart Contracts Empirical Study and Code Metrics*. IEEE, 2019.
44. E. Nicolette, *6G Will Need to Leverage Blockchain, Machine Learning, and AI: Research*. IEEE Generation, 2020.
45. T. Twenhoven, and M. Petersen, *Impact and Beneficiaries of Blockchain in Logistics*. Hamburg International Conference of Logistics, 2019.
46. P. Dutta, T. M. Choi, S. Somani, and R. Butala, "Blockchain technology in supply chain operations: Applications, challenges and research opportunities," *Transportation Research. Part E, Logistics and Transportation Review*, vol. 142, p. 102067, 2020. https://doi.org/10.1016/j.tre.2020.102067.

Chapter 4

Blockchain and Artificial Intelligence–Based Radio Access Network Slicing for 6G Networks

Ying Loong Lee,[1] Allyson Gek Hong Sim,[2] Li-Chun Wang,[3] and Teong Chee Chuah[4]

CONTENTS

[1] Department of Electrical and Electronic Engineering, Lee Kong Chian Faculty of Engineering and Science, Universiti Tunku Abdul Rahman, Selangor, Malaysia
[2] Secure Mobile Networking (SEEMOO) Lab, Department of Computer Science, Technical University of Darmstadt, Darmstadt, Germany
[3] Department of Electrical and Computer Engineering, National Chiao Tung University, Hsinchu, Taiwan
[4] Faculty of Engineering, Multimedia University, Selangor, Malaysia
Corresponding author: leeyingl@utar.edu.my

DOI: 10.1201/9781003264392-4

4.1 INTRODUCTION

New applications are emerging fast and will drive the next-generation wireless networks, the sixth generation (6G). These emerging applications are envisaged to embrace five senses—sight, hearing, smell, taste, and touch—which give rise to more stringent requirements compared to the fifth-generation (5G) key driving applications: enhanced mobile broadband (eMBB), ultrareliable low-latency communications (URLLC), and massive machine-type communications (mMTC). Also, Industry 4.0 has driven many industrial applications such as autonomous robots, automated manufacturing, and the industrial Internet of Things (IoT) that require massive data transfer and processing. Besides that, similar applications also emerge for smart cities, e.g., autonomous driving vehicles and unmanned aerial vehicles (UAVs). Furthermore, ubiquitous connectivity beyond terrestrial domains such as air, space, and undersea are envisaged beyond the 5G era. The next-generation wireless networks need to be extremely flexible to meet the requirements of the preceding applications and to be capable of supporting various applications with highly diverse requirements simultaneously without interruption to the quality of service (QoS).

Network slicing, which first emerged as one of the key features of 5G wireless communications, has been envisioned to be the main key enabling technology for next-generation wireless communications [1]. The main advantage of network slicing is its ability to customize network settings and resources for a particular use -case or application in the form of a virtual *network slice*. Different network slices can be generated to accommodate various applications with completely different requirements simultaneously while maintaining a certain extent of isolation. Other network slices do not easily exploit resources assigned to each network slice.

Although network slicing has been researched and developed for 5G networks, most of the proposed network slicing mechanisms lack intelligence, which is crucial when a large number of parameters and factors are involved. That said, artificial intelligence (AI) is of utmost importance for future 6G networks as the 6G key driving applications involve numerous parameters (e.g., latency, packet loss, data rate). On the other hand, resource isolation among network slices, which is vital for guaranteed QoS for each network slice, remains an open challenge. Blockchain, a distributed ledger technology, has recently been regarded as a promising solution for secure and reliable network slicing. It can ensure that network resources are fairly shared among network slices.

4.1.1 Motivations and Contributions

Given the emergence of many new use-cases with diverse requirements demanding various types of resources (e.g., radio channels, storage, and computational capacity), fully flexible and customizable network configurations are crucial for future wireless networks. Network slicing will certainly be one of the main enablers for future wireless communications. Nevertheless, in-depth research in this area remains critical because the current advances in network slicing are still far from enabling full RAN flexibility. Hence, this chapter aims to address this issue by highlighting the critical aspects of network slicing that remain underexplored. To this end, the role of AI and blockchain in empowering future RAN slicing will be highlighted.

The main contributions of this chapter can be summarized as follows:

1. Critical aspects of network slicing; namely *(i) RAN function virtualization and placement* and *(ii) RAN resource management and isolation* are discussed. An example 6G scenario is given to demonstrate the importance of these aspects.
2. The importance of AI as an enabler for 6G RAN slicing is described. In particular, potential AI applications in the *(i) network slice instance layer, (ii) resource layer*, and *(iii) RAN edge* are presented.
3. The role of blockchain in 6G RAN slicing is elucidated. Complementary implementation of both AI and blockchain for RAN slicing is also highlighted in a case study.

4.1.2 Chapter Organization

In this chapter, insights into the application of AI and blockchain for network slicing in 6G networks are provided. In particular, this chapter focuses on RAN, which plays a major role in wireless communications. First, Section 2 explores and identifies several key driving applications of 6G. Then Section 3 provides an overview of network slicing, including core network slicing and

RAN slicing. After that, Section 4 discusses RAN slicing and its role as the key enabler for 6G. Next the application of AI for RAN slicing is described in Section 5. Following that, blockchain is introduced along with its roles and recent advances on RAN slicing in Section 6. Then a case study is presented in Section 7 to investigate the characteristics and challenges of integration of both AI and blockchain for RAN slicing. Finally, the chapter is concluded in Section 8 with several remarks.

4.2 6G KEY DRIVERS

Following the rollout and large-scale deployment of 5G in 2019 and 2020, 6G targets not only to further enhance the support for the existing 5G services but it also to reinforce applications empowered by artificial intelligence (AI) that disrupt the original 5G goals.[1] Specifically, AI-driven applications are expected to transform the way of life of all humankind through digitalization, automation, internetworking, and network intelligence.

In this section, we discuss several promising AI-driven key applications, namely 5D applications, supersmart cities, and industries, as well as nonterrestrial communications (see Figure 4.1).

4.2.1 5D Applications

5D applications are intended to feature services and communications involving five-dimensional (5D) senses (i.e., sight, hearing, smell, touch, and taste). These applications include but are not limited to wireless brain–computer interactions, telemedicine, and extended reality that encompasses augmented reality (AR), virtual reality (VR), and mixed reality (MR) [2]. These applications encompass high-precision tactile, haptic, and holographic

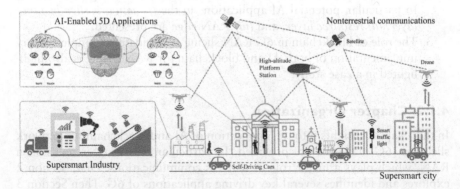

Figure 4.1 6G key drivers.

communications to provide experiences that are very close to reality for the end users.

Realizing that 5D applications pose unprecedented challenges for the existing wireless technologies from a multitude of aspects. In particular, these applications not only demand outstanding legacy performance (throughput, delay, spectral and energy efficiency, and reliability), they will also demand new performance metrics (context awareness, precise sensing, delay jitter, control, computing functionalities, a high degree of network intelligence, the convergence of communication and mobility support). For instance, while the introduction of high-frequency millimeter-wave communications for 5G networks offers Gbps rates, the surge in XR applications demands Tbps rates, which quickly depletes the millimeter-wave spectrum. Calling for the need to exploit teraHertz frequencies [3], the teraHertz band provides precise sensing for telemedicine sectors such as sensor implantation for extreme precision healthcare monitoring.

Nevertheless, achieving Tbps rates in scenarios with high mobility nodes (e.g., vehicles, UAVs, high-speed trains) at such high frequencies is extremely difficult. Such an issue will be further exacerbated for 5D applications as they demand a multitude of performance requirements that must be simultaneously guaranteed. Furthermore, realizing XR requires capturing all sensory inputs and delivering low-latency performance for data rate–intensive XR applications. Fulfilling such requirements requires an integrated design from both engineering (e.g., computing, storage, and wireless communication) and perceptual (e.g., human senses, physiology, and cognition) perspectives.

4.2.2 Supersmart Cities and Industries

Supersmart cities and industries correspond to the widespread use of self-controlled devices with an extremely high level of autonomy in cities (e.g., cars, public transport, transportation systems, street lamps, traffic lights) and industries (e.g., robots, production lines, UAVs). The autonomy of these devices goes beyond the typical capability of wireless devices that only executes restricted control such as local data collection. Specifically, these devices' decision processes may involve interconnecting devices in proximity or remote locations to make optimal local and global decisions. For instance, the maneuvering decision of a self-driving car must account for critical information such as traffic conditions, the mobility patterns of other vehicles, and congestion. The exchange of such information provides precise control of a device, but it requires stringent guaranteed requirements across the rate reliability–latency spectrum. In particular, the basic requirements are ultrahigh data rate (1 Tbps), ultralow latency (which is typically less than 1 ms), and 99.9999% reliability. These requirements are beyond the 5G performance limits of 10 Gbps, 10 ms, and 99.999%, respectively [4, 5].

Apart from achieving excellent performance, other critical issues of autonomous wireless systems are sustainability and scalability. From the sustainability perspective, autonomous devices should operate at low powers with self-recharging capability, for example, through wireless power transfer. Sustainability is crucial to maintain seamless communication and undisrupted operations to avoid undesirable outages, leading to production downtime and accidents in commercial and industrial premises, respectively. From the scalability perspective, the backbone (e.g., networking) of an autonomous system is expected to support massive connectivity and interconnectivity of diverse autonomy-enabling components (e.g., high-precision camera, LiDAR, RADAR, caches, computing elements, controllers, sensors) delivering services with heterogeneous requirements.

Because of the complex and computationally intensive operations needed to materialize autonomous systems for supersmart cities and industries and the unpredictability of wireless connectivity and dynamicity of communication environments, the involvement of machine learning and AI [6] becomes inevitable in guaranteeing smooth operational excellence.

4.2.3 Nonterrestrial Communications

One of the persistent challenges since the emergence of wireless communication is falling short of providing reliable wireless connectivity in remote/rural areas, overcrowded and disastrous zones, as well as undersea environments [7]. In light of this shortage, one of the 6G research focuses is developing nonterrestrial networks (NTNs) to support ubiquitous wireless connectivity and provide robust communications to those challenging environments. In particular, NTNs complement the existing terrestrial infrastructure by incorporating UAVs, high-altitude platform stations, and satellites into its ecosystem [8]. An apparent benefit of NTNs is the use of UAVs to provide either additional bandwidth in areas with a sudden surge in capacity demand. Such capacity surge may result from events at points of interest or the need for temporary infrastructure-less support (i.e., ad hoc networks) in critical scenarios with connectivity outages.

While beneficial, such a network requires meticulous coordination among UAVs and the existing ground stations to achieve optimal performance, resulting in a performance trade-off. For instance, UAVs could be flexibly placed to provide optimal capacity, but they have limited power. The advancement in undersea communication provides noninvasive marine life monitoring and underwater oil exploration and is also capable of carrying information over long distances using acoustic waves. Furthermore, autonomous undersea vehicles potentially facilitate efficient underwater aircraft or ship recovery.

As these applications of NTNs are typically challenging, their performance requirements are more difficult to guarantee. That said, many open issues are spanning multiple aspects that needed to be addressed. For instance, reconfigurable wireless and optical communications and cross-layer optimization are necessary to ensure network convergence in diverse and harsh communication environments.

4.3 NETWORK SLICING

In this section, an introduction to and historical background of network slicing is provided. An overview of core network slicing and that of radio access network (RAN) slicing are also given.

4.3.1 Background

Early network slicing is envisioned to achieve efficient infrastructural sharing between mobile network operators (MNOs). By leveraging network function virtualization (NFV) and software-defined networking (SDN) technologies, network slices have further evolved and can be realized by "slicing" the physical network resources into several *network slices*. These network slices are logical virtual networks with each customized for a particular service or MNO [9]. The service or MNO is known as the owner of the network slice and is also called the *tenant* of the network.

Network slicing introduces the following business opportunities [9, 10]:

- MNOs can act as infrastructure providers (InPs) by leasing some portion of their infrastructures and resources in the form of network slices to mobile virtual network operators (MVNOs). These MVNOs can be MNOs that seek coverage or capacity expansion in a given geographical area or those who do not own physical network infrastructure.
- InPs can also provide on-demand service guarantees in the form of network slices to third-party service providers such as over-the-top (OTT) service providers (e.g., video streaming) and vertical industries (e.g., e-health, surveillance, and automotive). In other words, the physical network infrastructure can be shared among not only MNOs but also MVNOs and third-party service providers. Such network sharing is known as multitenant network slicing, where tenants often refer to the owners of network slices, i.e., MNOs, MVNOs, OTT service providers, and verticals. This multitenant network slicing paradigm not only reduces CAPEX and OPEX but also optimizes resource utilization of the physical network infrastructure.

Network slicing has been proposed for 5G systems to allow resource customization for three different fundamental pillars [11]:

- *eMBB*: Applications and services that are characterized by high data rate requirements.
- *URLLC*: Applications and services where low latency, ultrahigh reliability, and high precision accuracy are crucial.
- *mMTC*: Applications and services involving large numbers of Internet of Things (IoT) devices.

For example, an eMBB slice will be provisioned with more bandwidth to meet the high data rate requirements, whereas an mMTC slice will receive less bandwidth for low-rate IoT applications.

In general, network slicing can be classified into core network slicing and RAN slicing. Core network slicing mainly deals with virtualization of core network elements to form core network slices [12], whereas RAN slicing deals with virtualizing RAN elements and RAN resource management [9]. Figure 4.2 summarizes core network slicing and RAN slicing.

4.3.2 Core Network Slicing

NFV allows various core network functions to be virtualized, so-called virtual network functions (VNFs) and grouped under an SDN paradigm to form core network slices. Each could be tailored to meet certain requirements of a particular service. These VNFs include packet forwarding functions, mobility

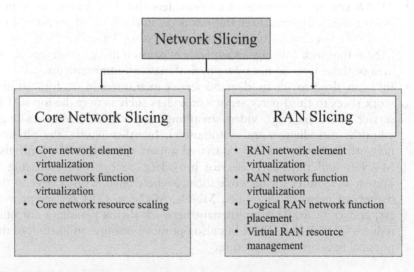

Figure 4.2 Example 6G network supporting emerging applications.

management, session management, policy enforcement, etc., which can be flexibly and dynamically linked or released according to the needs of different network slices supporting different MNOs or services [13]. The main advantage of core network slicing is that it can simplify the protocol stack of a core network slice according to the specifications and requirements of the supported service.

4.3.3 Radio Access Network (RAN) Slicing

RAN slicing encompasses virtualization of the RAN protocol stack and functions, such as radio resource control (RRC), radio link control (RLC) and medium access control (MAC) [9], and management of RAN resources such as baseband computing resources, channel bandwidth, transmission power, and mobile base stations (BSs). RAN slicing enables flexible RAN functions placement for each RAN slice according to its requirement. For instance, real-time functions such as MAC could be placed closer to the radio front end for time-critical RAN slices, whereas non-time-critical network slices can have most of the RAN functions centralized at a radio cloud [14]. RAN resources such as spectrum and power are also allocated according to the service requirement of each RAN slice while maintaining a satisfactory level of resource isolation.

4.4 RAN SLICING AS KEY ENABLER FOR 6G

With the various emerging applications mentioned in Section 2, wireless communications beyond 5G must be highly flexible in resource allocation to effectively meet the dynamic requirements of the applications. In this regard, RAN slicing plays a vital role, especially in quality of service (QoS) provisioning for time-critical services. This section discusses several aspects of RAN slicing, which are crucial to enable 6G to support future applications with highly diverse requirements, followed by a discussion on an exemplary 6G scenario.

4.4.1 Key Aspects and Requirements

In general, RAN slicing mainly involves virtualization and placement of RAN functions and RAN resource management and isolation. These aspects are of paramount importance to support emerging 6G use-cases.

4.4.1.1 RAN Function Virtualization and Placement

Cloud RANs [15] have been introduced as a new architecture that allows centralization of some RAN functions at a cloud server, also known as the baseband unit (BBU) pool or the RAN cloud. In contrast, some other RAN functions are decentralized to the RAN front end (e.g., BSs). With NFV technology, logical RAN functions (e.g., medium access control [MAC] scheduling, hybrid automatic repeat request [HARQ], radio link control [RLC], radio resource control [RRC], etc.) can be flexibly placed between the RAN cloud and the RAN front end, depending on the application requirements of use-cases [14].

Under the network slicing framework, different logical RAN functions can be grouped and linked together to form multiple RAN slices, with each servicing a different application. The grouping of different logical RAN functions depends on the specific requirements of the RAN slice. For example, more control plane (CP) functions than user-plane (UP) functions are required in a RAN slice that serves IoT applications since they involve more control signaling than data transfer. Besides that, the logical RAN function placement between the RAN cloud and the RAN front end can be configured according to the slice-specific requirements. For instance, logical real-time MAC scheduling should be placed in the radio front end for RAN slices supporting time-critical use-cases such as autonomous driving.

4.4.1.2 RAN Resource Management and Isolation

5G mobile cellular systems support a flexible frame structure based on scalable orthogonal division frequency multiplexing (OFDM) numerology [11]. Under this flexible frame structure, radio resources (also known as the resource blocks [RBs]) of different time-frequency sizes are available, which can be chosen according to the requirements of the use-cases. For example, RBs with wider subcarrier spacings are suitable for low-latency high-bandwidth applications, whereas those with narrower subcarrier spacings can be used for low-rate massive IoT applications. Often, RB selection is accompanied by transmission power allocation to increase the signal-to- interference-plus-noise ratio (SINR), thus improving achievable capacity. Since transmission power and RBs are limited, resource management among RAN slices has become vital for slice provisioning. Fair resource allocation among RAN slices can be achieved via a service-level agreement (SLA) established in advance. This SLA determines the priority of RAN slices in acquiring resources, which could be based on QoS requirements and service types. Nevertheless, several studies have proposed dynamically providing resources to RAN slices according to their service types and QoS requirements to achieve high resource utilization. However, it has been highlighted that resource isolation has to be kept to some degree to guarantee resources and QoS provisioning for each RAN slice [9]. For example, the resources for RAN slices supporting low-latency applications should be highly isolated from the other RAN slices, which are not resource intensive.

4.4.2 RAN Slicing for 6G

6G networks will need to support numerous applications simultaneously. Moreover, future applications will impose more diverse QoS requirements, and some are even more resource intensive than 5G, as described in Section 2. RAN slicing is certainly a key technology toward enabling 6G networks to support these emerging applications. Based on the cloud RAN architecture, Figure 4.3 shows an example RAN slicing scenario supporting various applications. In particular, three RAN slices are instantiated, each servicing a particular use-case. The RAN slice 1 of Figure 4.3 supports latency-critical and reliability-critical applications such as connected autonomous vehicles and UAVs. Most of the logical RAN functions are placed at the RAN front end because such placement minimizes communication and processing delays, thus improving latency reduction and reliability. In addition, a higher OFDM numerology, with a large subcarrier spacing and shorter time slot, is needed to guarantee the latency and reliability requirements.

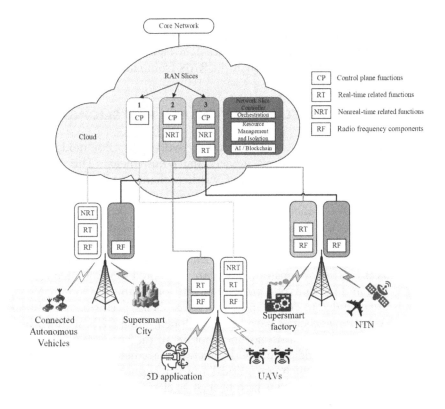

Figure 4.3 Example 6G network supporting emerging applications.

The RAN slice 2 of Figure 4.3 supports latency-critical, bandwidth-intensive use-cases, such as 5D applications and supersmart factory. This RAN slice has real-time network functions (e.g., MAC scheduling) placed at the RAN front end to minimize latency. On top of that, this RAN slice needs to be allocated more radio resources from an OFDM numerology with a larger subcarrier spacing to support the high-bandwidth requirement.

The RAN slice 3 of Figure 4.3 supports supersmart city applications and airplane and satellite communications, in which the latency requirement is not critical. It is observed that the RAN slice has most of its RAN functions placed at the RAN cloud, leaving only RF functions at the RAN front end since the serviced use-case is delay tolerant. In addition, supersmart city applications such as low-rate IoT devices only require an OFDM numerology with smaller subcarrier spacing. For NTN communications, an intermediate OFDM numerology may be required.

In Figure 4.3, resource management and isolation for RAN slicing are determined at the RAN cloud, whereby resources are allocated, shared, and isolated at both the RAN cloud and front end. The RAN cloud and front end may have their respective dynamic resource sharing mechanisms to enhance resource utilization, but some resource isolation is in place. For latency and reliability-critical RAN slices (e.g., slice 1 in Figure 4.3), a higher degree of resource isolation may be needed at the RAN front end.

4.5 ARTIFICIAL INTELLIGENCE (AI) FOR RAN SLICING

Flexible RAN slicing is crucial for 6G networks to accommodate various emerging applications' various services and requirements. This challenge is exacerbated by the rapidly varying wireless channel, which places fundamental limits on RAN performance. Various studies have been carried out to address the RAN slicing problems via nonconvex programming methods [16, 18]. The RAN slicing problem is often formulated as a mathematical optimization problem that aims to achieve certain objectives such as maximizing throughput, maximizing energy efficiency [16, 18] and optimizing fairness [17], subject to a set of constraints that are attributed to the limited resources such as RBs and transmission power. In some formulations, QoS-related constraints are included to ensure minimum RAN slicing performance. Although many RAN slicing schemes have been developed, these RAN slicing schemes often are suboptimal and do not incorporate all the key aspects (e.g., frame structure, spectrum, logical network function placement, and radio resource management, etc.), which are critical toward supporting future emerging applications. Moreover, other factors such as channel conditions, carrier frequencies, user locations, and densities have to be taken into account to guarantee network performance, rendering decision making of RAN slicing a very challenging task. AI has been regarded as a promising candidate to achieve efficient and optimal decision making for RAN slicing. AI has witnessed many successes in various other fields such as robotic automation, self-driving cars, image recognition, etc. The main feature of AI is that it can take in many input parameters to produce an appropriate output or decision without manual human intervention. In what follows, we discuss how AI can be applied to different levels of RAN slicing.

4.5.1 AI in the Network Slice Instance Layer

While the 5G system adopts a service-based architecture to achieve "connected things," the 6G architecture is expected to be AI native by leveraging AI and machine learning to achieve "connected intelligence" [19]. Network slicing is a powerful technology in 5G for providing virtualization capability to create multiple logical networks over a common physical infrastructure, allowing network resources to be dynamically and elastically allocated to support a wide range of services with diverse QoS requirements for the isolated logical networks.

Advanced 6G use-cases and services place heavier demands on RAN performance than 5G in terms of latency, reliability, and efficiency in a more complex and ultraheterogeneous environment. Hence managing functions and resources at all network levels under network slicing will become a daunting task. By enabling machine learning and fast adaptation, a 6G network paradigm that embraces AI-based network slicing will render service delivery more agile and efficient, taking 6G to a whole new level, namely toward an "AI native" network.

As depicted in Figure 4.4, the network slicing concept defined by the Next Generation Mobile Networks (NGMN) Alliance comprises three layers [20]: (1) service instance layer, (2) network slice instance layer, and (3) resource layer. The service instance layer represents the services to be provided by the network operator or by third parties. A network operator uses the blueprint of a network slice to establish a network slice instance, which supports the features needed by a service instance. A network slice instance may also be shared across multiple service instances provided by the network operator. The resource layer consists of physical resources such as network infrastructure and logical resources such as RBs.

AI should be deployed and trained at different layers of the network slice with the assistance of an intelligent orchestrator. For example, in the network slice instance layer, AI can be deployed to stitch together the network functions required to meet the network characteristics for a service instance. Machine learning algorithms can be used for predicting and estimating the traffic demand and mobility behaviors of each network slice instance. Based on current traffic conditions and the estimated future capacity, the InP can predict whether the SLA of a new tenant can be fulfilled, thus enabling slice preparation and admission/rejection and fronthaul capacity allocation. Then, by leveraging the predicted traffic arrival dynamics, elastic interslice resource distribution among the network slice instances can be performed using reinforcement learning to minimize under-/overprovisioning and to realize various intelligent power-saving features (e.g., sleep mode) for improved power efficiency. Moreover, AI can also be applied for slice customization, i.e., optimizing the placement of logical network functions, chaining multiple logical network functions, virtual network function placement, and logical network function scaling for each network slice instance according to the SLA and network conditions. For example, tenants running a supersmart city/factory may not need demanding mobility-related network functions. Thus network functions that involve mobility control can be scaled down accordingly in the corresponding network slice instances. For tenants serving mobile devices, higher-layer network functions such as handover can also be optimized with the controlled and predictable deployment of machine learning agents. For 5D and autonomous driving applications with stringent latency requirements, efficient AI-based network function placement algorithms can be developed to facilitate QoS enforcement, for example, by placing real-time network functions such as dynamic medium access control (MAC) scheduling closer to the edge to minimize latency.

4.5.2 AI in the Resource Layer

The degree of control available in a 6G RAN will be unprecedented. As in previous generations, there will be parameters that determine the characteristics of the physical radio interface, including allocation of spectrum, transmit power, coding, OFDM numerology, beamforming, antenna orientation, etc. Early generations of mobile networks largely relied on manual or semi-autonomous tuning. The extreme range of parameters to be controlled in 6G networks will drive the need for AI-driven automation, which is essential to achieve a zero-touch optimization at a higher level of operational efficiency.

An AI-based data processing platform can process the huge volumes of live network data generated by a 6G network to improve functionalities such as anomaly detection and predictive maintenance at the resource layer. Site interventions can be reduced through automated site inspections with AI-controlled drones. AI-based antenna tilting can be deployed to optimize coverage and capacity by adjusting base station antennas' tilt based on the dynamics of the network environment. Unlike conventional rule-based antenna tilting, AI techniques enable self-improving learning from feedback through network performance data. Using reinforcement learning, an agent can be trained to dynamically control the tilting of multiple base stations to improve the signal quality in a cell while reducing interference to neighboring

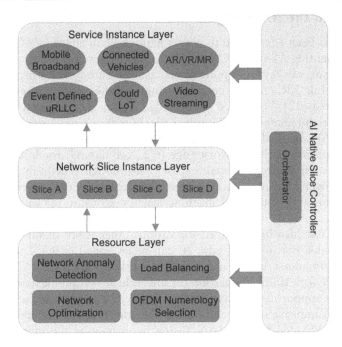

Figure 4.4 AI-based network slicing.

cells in response to real-time traffic and mobility patterns. This results in an overall network performance improvement while reducing operational costs. Moreover, deep learning algorithms can be adopted for interference coordination, coverage and capacity optimization, and load balancing to optimize RAN performance [21]. These algorithms can be centralized, distributed to the edge, or both.

Because of the ultrahigh heterogeneity of 6G networks, the scalable OFDM numerology introduced in 5G needs to be expanded to include more options on subcarrier spacing and slot duration to accommodate a wider range of factors including service type, QoS requirements (e.g., throughput, packet loss, latency), carrier frequency, and cell size. To create different RAN slice instances for tenants with different combinations of these factors, the 6G physical layer needs to implement a dynamic numerology selection algorithm. By exploiting the terrain map, location, and mobility data of the devices, the underlying channel parameters can be gleaned via big data analytics. Then, given this knowledge and other service-related information, an AI system can be developed to perform dynamic numerology selection, for example, on a per-cell or per-network slice instance basis.

4.5.3 AI in the RAN Edge

Edge AI has been envisioned to play a major role in enabling 6G networks. By implementing AI at the RAN edge, i.e., at the network nodes such as cellular BSs, several key 6G applications such as autonomous driving, smart homes, mobile XR, environmental sensing, and the smart factory can better be supported [22]. For RAN slicing, edge AI can be leveraged to improve intraslice resource management, particularly at the network edge. For example, edge AI can facilitate the processing of the real-time-related network functions placed at the network edge (e.g., cellular BS) for latency-critical use-cases, such as decision making for real-time RB scheduling and transmission power allocation among users. AI techniques such as evolutionary optimization and distributed machine learning are suitable for performing such tasks.

Edge AI can also be leveraged for various other RAN slicing use-cases in different tasks. For instance, edge AI can be leveraged in RAN slices, which are for autonomous driving use-cases, to enable dynamic spectrum access to mitigate network traffic congestion for intervehicle communications [22]. For RAN slices supporting XR applications, edge AI can provide computational offload and improve end-to-end latency. In addition, edge AI can cooperate with the RAN cloud AI in RAN slicing by estimating the number of edge resources, such as RBs, for communications and computational resources for the radio front end required by each BS for each RAN slice, based on the channel conditions and user equipment (UE) states, as well as the status of traffic and resource availability for each RAN slice at the RAN edge.

AI techniques such as Q-learning [23] and deep reinforcement learning [24] can be used to determine proper actions, i.e., appropriate amounts of resources reserved for RAN slices at the RAN edge, by estimating the *reward* gained from such actions based on the channel, UE, resource and traffic states of each RAN slice. Figure 4.5 shows an application of edge AI for RAN slicing. Moreover, by determining the number of radio resources that can be shared among the RAN slices and the minimum amount of radio resources reserved for each RAN slice, edge AI can coordinate resource sharing and isolation among multiple RAN slices at the same edge node.

4.5.4 Advances in AI for RAN Slicing

Several AI techniques have been investigated to implement AI-based RAN slicing. In particular, deep learning has been leveraged in Khodapanah et al. [25] to address a radio resource management problem for RAN slicing. An artificial neural network is developed to fulfill all the QoS requirements (e.g., data rates, drop rates, and admission rates) of each slice.

Compared to deep learning, reinforcement learning has been studied more intensively for RAN slicing. In Dandachi et al. [26], reinforcement

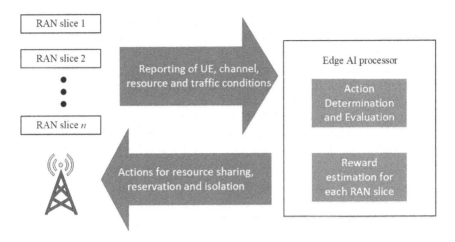

Figure 4.5 Application of edge AI for RAN slicing.

learning is used to solve the congestion and admission control problem that aims to achieve a trade-off between the number of accepted RAN slices and the rejection rate of network slice requests. In another study [27], a RAN slicing scheme based on offline reinforcement learning is proposed to assign radio resources to RAN slices so that resource utilization is maximized while guaranteeing the traffic requirements of each RAN slice. The study in Xiang et al. [28] has developed two reinforcement learning techniques for fog RAN slicing to address the system power minimization problem under diverse user demands and limited computational resources in different network slice instances. Interestingly, a group of researchers has combined both reinforcement learning and deep learning for RAN slicing, where the former is used to perform small time-scale resource allocation. In contrast, the latter is to perform large time-scale resource allocation so that the resource utilization is optimized while guaranteeing a certain degree of RAN slice isolation [29].

Recently, deep reinforcement learning has gained much attention for RAN slicing due to the use of neural networks in addressing the memory problem in reinforcement learning. In Van Huynh et al. [30], a deep double Q-learning algorithm with deep dueling is developed to allocate radio, computational, and storage resources for each RAN slice. Deep reinforcement learning has also been used in Abiko et al. [31] to enable flexible radio resource allocation for RAN slicing, which aims to meet network slice requirements and enhance resource utilization. In another study [32], the authors proposed a deep reinforcement learning technique known as generative adversarial network-based deep distributional Q-learning to allocate bandwidth to different RAN slices according to their bandwidth demands. In Xiang et al. [33], a deep Q-network is used to address the content caching and mode selection problem for fog RAN slicing under the conditions of limited resources and diverse user demands.

Table 4.1 shows the comparison of the preceding AI-based RAN slicing schemes. Reinforcement learning and deep reinforcement learning have gained more interest in RAN slicing research due to their inherent nature of model-free learning. On the other hand, most of the studies have well considered the QoS requirements of RAN slices, but RAN slice isolation is largely ignored. Also, virtual function placement has not been considered part of the slicing mechanism in the AI-based RAN slicing frameworks proposed in these studies.

Table 4.1 Comparison of AI-Based RAN Slicing Schemes

Study	AI technique	Objective	Slicing mechanism
[25]	Deep learning	Fulfill QoS requirements (e.g., data rates, dropping rates, and admission rates)	Radio resource management
[26]	Reinforcement learning	Maximize the number of accepted RAN slices while minimizing the rejection rate of network slice requests	Congestion and admission control
[27]	Reinforcement learning	Maximize resource utilization and guarantee RAN slice traffic requirements	Radio resource allocation
[28]	Reinforcement learning	Minimize system power consumption	Computational resource allocation under diverse user demands
[29]	Reinforcement learning and deep learning	Maximize resource utilization while guaranteeing RAN slice isolation	Radio resource allocation
[30]	Deep reinforcement learning	Maximize a specific user-defined reward	Radio, computational and storage resource allocation
[31]	Deep reinforcement learning	Maximize resource utilization and fulfill network slice requirements	Radio resource allocation
[32]	Deep reinforcement learning	Fulfill bandwidth demands of RAN slices	Bandwidth allocation
[33]	Deep reinforcement learning	Maximize the cache hit ratio while minimizing residual bits of RAN slices	Caching and mode selection

4.6 BLOCKCHAIN FOR RAN SLICING

In this section, the application of blockchain in RAN slicing for 6G networks is discussed. An introduction to blockchain will be given, followed by a discussion on blockchain application for RAN slice resource brokering and potential integration of AI and blockchain for 6G RAN slicing.

4.6.1 Background of Blockchain

Blockchain is a distributed public ledger technology that has been used in the famous cryptocurrency known as Bitcoin. This technology provides a decentralized, secure, and immutable ledger, allowing transactions to be made without the involvement of a central entity [34]. In a blockchain, each block is a record that contains data (e.g., amount of transaction), a unique hash that identifies the block, and a reference to the previous block in the blockchain. Since a blockchain is a peer-to-peer network whereby the peers (or also known as nodes) collectively validate the information in the new block, a newly created block will be broadcast to all the participating nodes in the blockchain to be validated. The validation protocol is also called the consensus protocol of the blockchain, which prevents *double-spending* [35] and protects the data in each block from being tampered with. Moreover, blockchain can protect data integrity and enhance the utilization of power and resources for a mobile network such as IoT.

4.6.2 Role of Blockchain in 6G RAN Slicing

The idea of RAN slice brokering has emerged as a means for MVNOs to request and release the required amount of RAN resources from the InP [36]. In this regard, blockchain has been deemed to provide an effective brokering mechanism for RAN slicing. Depending on the channel conditions, the wireless network can divide these resources orthogonally among the RAN slices; i.e., no common resources are shared between RAN slices or allow some RAN slices to share certain RAN resources such as spectrum under the scenario where interference is minimum, and the QoS is not jeopardized. In such a situation, blockchain can ensure that the resources divided or shared between RAN slices are sufficient and are fairly distributed among RAN slices by storing the information of resources allocated to the RAN slices (e.g., types of resources, amounts of resources allocated, resource index numbers) [37]. Besides that, performance targets such as achieved throughput and latency can also be recorded in the blockchain.

With blockchain, SLA fulfillment can be checked and ensured for the RAN slices via *smart contracts*, which are executable codes developed using a specific programming language and accessible by all the blockchain participants [34]. These smart contracts can be encoded with SLA-aware resource allocation schemes and executed when the tenants make RAN slice requests [38] [39].

Blockchain can enforce trust and security in RAN slice brokering. Blockchain potentially provides a transparent and fair RAN slice brokering system that can avoid resource starvation, guarantee resource isolation, and prevent malicious MVNOs from requesting more resources than what is needed or using other RAN resource slices associated with other MVNOs [40]. In this regard, the *consensus algorithm* plays an important role in validating resource allocation decisions.

4.6.3 Advances in Blockchain for RAN Slicing

Blockchain has the advantage of guaranteeing resources for each RAN slice and therefore is a means for achieving resource isolation among RAN slices. Despite that, only a handful of studies have attempted to investigate blockchain applications in RAN slicing. In one study [35], the authors proposed a blockchain-based RAN slicing architecture where each RAN slice is composed of RAN resources such as spectrum and transmit power. The proposed architecture consists of wireless service and resource providers (e.g., cellular and WiFi) that sublease time-frequency channels, services that process data about resource sharing, and block managers maintaining the blockchain. Each transaction in a block carries information about bandwidth allocation, transmit power level in the channel, data rate, etc., which the tenants use to serve their respective users. A transaction in the blockchain is made when:

1. A wireless resource provider subleases RAN resources to a tenant (i.e., the source is the provider, and the recipient is the tenant); or
2. A tenant releases the leased RAN resources back to the wireless service and resource provider (i.e., the source is the tenant, and the recipient is the wireless service and resource provider).

The proposed blockchain-based RAN slicing architecture allows each block to store a set of transactions, consisting of information about the RAN resource allocation used by the tenants. All participants in the blockchain (i.e., the wireless service and resource providers) will cooperatively create and verify new transactions that are to be added to the blockchain. A consensus algorithm known as proof of wireless resources (PoWR) is developed to generate a new block for storing new transactions and verifying RAN resources' availability by the wireless resource providers. Also, PoWR checks whether the new transactions follow the SLA established between wireless resource providers and tenants. This prevents the same RAN resources from being spent multiple times by multiple tenants, as such spending could cause inter-ference between tenants that will lead to SLA violations.

On the other hand, some authors have investigated blockchain with smart contracts for RAN resource brokering among RAN slices. For instance, the study in [38] develops a blockchain-based network slicing framework where tenants can auction their previously obtained resources among each other. Consider the fact that the requirement of each RAN slice may vary in time. Some RAN slices may require more communications resources (e.g., RBs), while others may require more computational resources (for signal processing) at a certain point in time. The blockchain-based network slicing framework allows these two RAN slices to auction and exchange their different resources via smart contracts to meet their individual needs. Once resources are auc-tioned, the transactions of resources are validated using a consensus algorithm by all the participants in the blockchain.

Another study [41] also applies blockchain to enable resource brokering among network slices but with a wider scope, i.e., core network and RAN resource brokering. In this study, when a tenant requests a network slice, it is published as a small smart contract in a permission-less blockchain where every participant of the blockchain can add new transactions to the block-chain. The resource providers then respond by publishing the costs of the resources requested by the contracts for tenants who are requesting a net-work slice. Once the tenant selects a contract, the network slice deployment are recorded in a permissioned blockchain where the blockchain participants cannot freely add new transactions to the blockchain. The slice provider, i.e., the entity that manages network slices, monitor the resources in each network slice and verify that the resource providers fulfill the SLA of the network slice.

Table 4.2 summarizes the comparison of the preceding blockchain-based RAN slicing schemes. It shows that blockchain has mainly been studied for resource brokering and to maintain the SLAs established among the RAN slices. Nevertheless, these studies have not considered the more complex use-cases such as those for smart cities requiring very high connection densities.

4.7 CASE STUDY ON INTEGRATION OF AI AND BLOCKCHAIN IN RAN SLICING

Although blockchain has been studied for 5G RAN slicing, these studies do not focus on the emerging 6G use-case requirements, which are more strin-gent than those of 5G. The role of blockchain in 6G RAN slicing and QoS provisioning is expected to grow further for achieving resource guarantee and isolation of RAN slices. On the other hand, AI has been deemed promising to make future networks fully automated and flexible. It is capable of taking into account a large number of parameters for decision making in RAN resource

Table 4.2 Comparison of Blockchain-Based RAN Slicing Schemes

Study	Objective	Blockchain application
[35]	Maintain RAN resource isolation, maintain SLA, and prevent interference between tenants in the use of spectrum	Store RAN resource allocation information, verify RAN resource availability, and check for SLA violations via the consensus algorithm
[38]	Enable exchanges of radio and computational resources between RAN slices	Enable resource brokering between RAN slices, and check on resource transactions' validity and SLA through smart contracts and a consensus algorithm
[41]	Maintain SLA of each network slice	Enable core and RAN resource brokering among network slices, and check for SLA violations

slicing. Given the advantages of both AI and blockchain technologies, it is worth exploring the integration and complementary use of both technologies for 6G RAN slicing. Such integration is very effective and efficient for smart applications such as UAVs, automated customer services, and energy trading. In this section, a case study on such integration is presented with the following objectives:

- To investigate the benefits and characteristics of the complementary use of both AI and blockchain in RAN slicing for 6G
- To explore the possible implementation framework for such integration in RAN slicing
- To identify the open challenges for such integration

4.7.1 Characteristics of AI and Blockchain for RAN Slicing

Blockchain can potentially empower AI for RAN slicing. One particular study in Adhikari et al. [42] has investigated the application of machine learning for blockchain-based RAN slice brokering. This study exploits a linear regression–based machine learning technique to predict the total data rate requirement of each MVNO while using blockchain to minimize double-spending of RAN slices by InPs for multiple MVNOs. However, this study does not exploit the features of the blockchain, which can be used to contain the resource usage and performance history of each MVNO. This information is tamperproof and can potentially be used by AI, such as deep learning to make better decisions in RAN slicing.

Information such as amounts of RBs, placement of logical network functions and selected OFDM numerology, and user throughput and latency can be stored in the blockchain, potentially useful for predicting the number of resources required and suitable network configuration in each RAN slice in the future time instances. Moreover, the information can identify SLA violations via some consensus algorithm for constraint verifications and penalize InPs who have caused double-spending (see Figure 4.6). In summary, blockchain-empowered AI-based RAN slicing can provide the following characteristics:

- Reduced double-spending of RAN slices and resources by InPs for multiple MVNOs
- Guaranteed resource isolation among RAN slices
- Reduced SLA violations between RAN slices or MVNOs
- Guaranteed QoS performance among MVNOs
- Efficient and fair RAN slice and resource allocation among MVNOs

4.7.2 Implementation of AI and Blockchain for RAN Slicing

Figure 4.7 illustrates the potential implementation framework of AI and blockchain for RAN slicing, based on Adhikari et al. [42]. In this implementation, a permissioned consortium blockchain, where only invited InPs and MVNOs can maintain, verify, and add new blocks to the blockchain, is considered. Each MVNO and InP is equipped with a block manager that handles

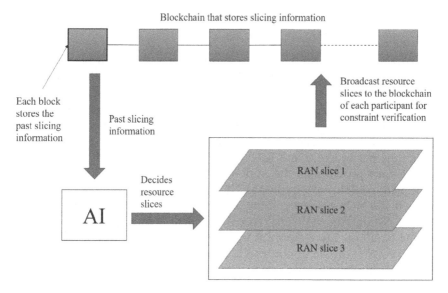

Figure 4.6 Combined application of AI and blockchain for RAN slicing.

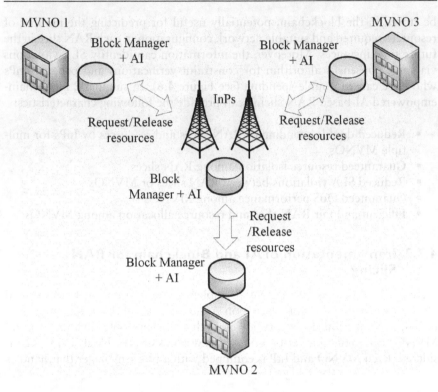

Figure 4.7 Implementation of AI and blockchain for RAN slicing.

the blockchain-based RAN slice brokering operations. Each block manager maintains a blockchain for each MVNO and InP to store the transactions containing the RAN slice information. The block managers can provide this information as the input to AI algorithms for predicting performance while validating the transactions of RAN slices to avoid double-spending. AI algorithms can be implemented either at the MVNO side or at the InP side to predict the performance and RAN slice requirements of each MVNO. With the outcomes provided by the AI, RAN slices with sufficient resources can be negotiated from the InPs to the MVNOs.

When MVNOs request resources from the InPs, smart contracts can be invoked to fulfill the MVNOs' requests via an SLA-aware resource allocation scheme based on game-theoretic approaches [39] and auction theory [38]. After executing the smart contracts, transactions for resource allocation are made and validated by all the participants in the blockchain via a consensus algorithm to ensure no double-spending of resources and SLA violations. Additionally, that SLA fulfillment can also be enforced through smart contracts. New transactions will be added as new blocks and broadcast to all the participants to update their respective blockchains in their respective block managers. On the other hand, the AI predictor in the block manager for MVNOs can collect the information from the blockchain to predict the future demand for resources. However, the complexity of continuously querying the blockchain for the resource slicing information is prohibitively high. Fortunately, training-based AI prediction techniques, such as deep learning using neural networks, can address this problem by allowing the block manager to automatically provide the data from a new block, once it is appended to its blockchain, to the neural network for training purposes. Thus this information will be fed to the AI predictor for training whenever a new block is generated. This process will help train the neural network without the need to access all the data from the blockchain when required to perform resource demand prediction, thus saving a huge computational burden. It is also noteworthy that the information from a new block is already validated; thus the information needed for training the AI technique is reliable and trustworthy, which will result in a more reliable resource demand prediction.

4.7.3 Open Challenges

Despite the benefits of integrating AI and blockchain in RAN slicing, several open issues and challenges should be addressed:

1. It is time-consuming to check for SLA violations and double-spending from the blockchain because all participants in the blockchain have to sign the block [42].
2. An efficient, accurate, and properly trained machine learning model, which is capable of considering a large number of input parameters such as types of services and QoS requirements, is required to determine the best network configuration and amounts of resources for all RAN slices.
3. An efficient implementation framework for integrating AI and block-chain into RAN slicing is essential. Centralized and decentralized implementation frameworks for such integration need to be further explored and analyzed.

4.8 CONCLUSION

Network slicing is undoubtedly a key enabling technology for flexible 6G networks for supporting various emerging use-cases, including 5D applications, supersmart city and industry, and nonterrestrial communications. In particular, RAN slicing plays a vital role in enabling 6G wireless communications to support these use-cases as the RAN places fundamental performance limits on QoS provisioning. In this chapter, an exemplary scenario of 6G RAN slicing has been envisioned and described. Furthermore, AI has been highlighted as a crucial element in RAN slicing for 6G networks, which can play major roles at the network instance layer, resource layer, and RAN edge. Also, some recent advances in AI-based RAN slicing have been reviewed and compared to identify open research gaps that need to be further addressed. Besides that, blockchain-based RAN slicing has been explored. In particular, the role of blockchain for RAN slicing in maintaining resource isolation, trust, and security has been emphasized. Moreover, recent advances on blockchain for RAN slicing have been reviewed and compared. Lastly, a case study on the potential combined applications of blockchain and AI in RAN slicing for 6G networks is investigated. From this case study, the characteristics and challenges of integrating AI and blockchain for 6G RAN slicing have been highlighted.

Acknowledgements This work has been carried out under the Fundamental Research Grant Scheme project FRGS/1/2019/ICT05/UTAR/02/1 and Fundamental Research Grant Scheme project FRGS/1/2019/ICT05/MMU/01/1 provided by the Ministry of Higher Education of Malaysia. Also, this work was in part supported by the DFG 6G-RATs project and by the LOEWE initiative (Hesse, Germany) within the emergenCITY center. Besides that, this work was supported by the Ministry of Science and Technology under Grant No. MOST 108–2634-F-009–006- and MOST 109–2634-F- 009–018- through Pervasive Artificial Intelligence Research (PAIR) Labs, Taiwan. This work was partially supported by the Higher Education Sprout Project of the National Chiao Tung University and the Ministry of Education, Taiwan.

NOTE

1 The 5G target applications are eMBB, uRLLC, and mMTC.

REFERENCES

1. Zhang, Z., Xiao, Y., Ma, Z., Xiao, M., Ding, Z., Lei, X., Karagiannidis, G.K., Fan, P.: 6G Wireless Networks: Vision, Requirements, Architecture, and Key Technologies. *IEEE Vehicular Technology Magazine* 14(3), 28–41 (2019).
2. Saad, W., Bennis, M., Chen, M.: A Vision of 6G Wireless Systems: Applications, Trends, Technologies, and Open Research Problems. *IEEE Network* 34(3), 134–142 (2020).
3. Giordani, M., Polese, M., Mezzavilla, M., Rangan, S., Zorzi, M.: Toward 6G Networks: Use Cases and Technologies. *IEEE Communications Magazine* 58(3), 55–61 (2020).
4. David, K., Berndt, H.: 6G Vision and Requirements: Is There Any Need for Beyond 5G? *IEEE Vehicular Technology Magazine* 13(3), 72–80 (2018).
5. Chen, Y., Liu, W., Niu, Z., Feng, Z., Hu, Q., Jiang, T.: Pervasive Intelligent Endogenous 6G Wireless Systems: Prospects, Theories and Key Technologies. *Digital Communications and Networks* 6(3), 312–320 (2020).

6. Yang, P., Xiao, Y., Xiao, M., Li, S.: 6G Wireless Communications: Vision and Potential Techniques. *IEEE Network* **33**(4), 70–75 (2019).
7. Giordani, M., Zorzi, M.: Non-Terrestrial Networks in the 6G Era: Challenges and Opportunities. *IEEE Network* **35**(2), 244–251 (2020).
8. Giordani, M., Zorzi, M.: Satellite Communication at Millimeter Waves: A Key Enabler of the 6G Era. In: *2020 International Conference on Computing. Net-working and Communications* (ICNC), pp. 383–388 (2020).
9. Afolabi, I., Taleb, T., Samdanis, K., Ksentini, A., Flinck, H.: Network Slicing and Softwarization: A Survey on Principles, Enabling Technologies, and Solutions. *IEEE Communications Surveys Tutorials* **20**(3), 2429–2453 (2018).
10. Chang, C., Nikaein, N., Arouk, O., Katsalis, K., Ksentini, A., Turletti, T., Sam-danis, K.: Slice Orchestration for Multi-Service Disaggregated Ultra-Dense RANs. *IEEE Communications Magazine* **56**(8), 70–77 (2018).
11. 3GPP: NR and NG-RAN Overall Description (2020). https://portal.3gpp.org/ desktopmodules/Specifications/ SpecificationDetails.aspx?specificationId=3191. Cited 21 July 2020.
12. Bagaa, M., Taleb, T., Laghrissi, A., Ksentini, A., Flinck, H.: Coalitional Game for the Creation of Efficient Virtual Core Network Slices in 5G Mobile Systems. *IEEE Journal on Selected Areas in Communications* **36**(3), 469–484 (2018).
13. Kaloxylos, A.: A Survey and an Analysis of Network Slicing in 5G Networks. *IEEE Communications Standards Magazine* **2**(1), 60–65 (2018).
14. Maeder, A., Ali, A., Bedekar, A., Cattoni, A.F., Chandramouli, D., Chan-drashekar, S., Du, L., Hesse, M., Sartori, C., Turtinen, S.: A Scalable and Flex-ible Radio Access Network Architecture for Fifth Generation Mobile Networks. *IEEE Communications Magazine* **54**(11), 16–23 (2016).
15. Checko, A., Christiansen, H.L., Yan, Y., Scolari, L., Kardaras, G., Berger, M.S., Dittmann, L.: Cloud RAN for Mobile Networks—A Technology Overview. *IEEE Communications Surveys Tutorials* **17**(1), 405–426 (2015).
16. Chang, Z., Han, Z., Ristaniemi, T.: Energy Efficient Optimization for Wire-less Virtualized Small Cell Networks with Large-Scale Multiple Antenna. *IEEE Transactions on Communications* **65**(4), 1696–1707 (2017).
17. Tun, Y.K., Tran, N.H., Ngo, D.T., Pandey, S.R., Han, Z., Hong, C.S.: Wire-less Network Slicing: Generalized Kelly Mechanism-Based Resource Allocation. *IEEE Journal on Selected Areas in Communications* **37**(8), 1794–1807 (2019).
18. Ho, T.M., Tran, N.H., Le, L.B., Han, Z., Kazmi, S.M.A., Hong, C.S.: Network Virtualization with Energy Efficiency Optimization for Wireless Heterogeneous Networks. *IEEE Transactions on Mobile Computing* **18**(10), 2386–2400 (2019).
19. Letaief, K.B., Chen, W., Shi, Y., Zhang, J., Zhang, Y.A.: The Roadmap to 6G: AI Empowered Wireless Networks. *IEEE Communications Magazine* **57**(8), 84–90 (2019).
20. NGMN Alliance: Description of Network Slicing Concept, NGMN 5G P1 Requirements and Architecture Work Stream End-to-End Architecture (2016). www.ngmn.org/wp- content/uploads/160113 NGMN Network Slicing v1 0.pdf. Cited 21 July 2020.
21. Luo, F.L.: *Machine Learning for Future Wireless Communications*. Wiley-IEEE Press (2019).
22. Peltonen, E., Bennis, M., Capobianco, M., Debbah, M., Ding, A., Gil-Castiñeira, F., Jurmu, M., Karvonen, T., Kelanti, M., Kliks, A., Leppänen, T., Lovén, L.,

Mikkonen, T., Rao, A., Samarakoon, S., Seppänen, K., Sroka, P., Tarkoma, S., Yang, T.: 6G White Paper on Edge Intelligence. *arXiv preprint arXiv:2004.14850* (2020).

23. Watkins, C., Dayan, P.: Technical Note: Q-Learning. *Machine Learning* 8, 279–292 (1992).

24. Xiong, Z., Zhang, Y., Niyato, D., Deng, R., Wang, P., Wang, L.: Deep Reinforcement Learning for Mobile 5G and Beyond: Fundamentals, Applications, and Challenges. *IEEE Vehicular Technology Magazine* 14(2), 44–52 (2019).

25. Khodapanah, B., Awada, A., Viering, I., Barreto, A., Simsek, M., Fettweis, G.: Framework for Slice-Aware Radio Resource Management Utilizing Artificial Neural Networks. *IEEE Access* 8, 174,972–174,987 (2020).

26. Dandachi, G., De Domenico, A., Hoang, D.T., Niyato, D.: An Artificial Intelligence Framework for Slice Deployment and Orchestration in 5G Networks. *IEEE Transactions on Cognitive Communications and Networking* 6(2), 858–871 (2020).

27. Albonda, H.D.R., Pérez-Romero, J.: An Efficient RAN Slicing Strategy for a Heterogeneous Network With eMBB and V2X Services. *IEEE Access* 7, 44, 771–44,782 (2019).

28. Xiang, H., Peng, M., Sun, Y., Yan, S.: Mode Selection and Resource Allocation in Sliced Fog Radio Access Networks: A Reinforcement Learning Approach. *IEEE Transactions on Vehicular Technology* 69(4), 4271–4284 (2020).

29. Yan, M., Feng, G., Zhou, J., Sun, Y., Liang, Y.: Intelligent Resource Scheduling for 5G Radio Access Network Slicing. *IEEE Transactions on Vehicular Technology* 68(8), 7691–7703 (2019).

30. Van Huynh, N., Thai Hoang, D., Nguyen, D.N., Dutkiewicz, E.: Optimal and Fast Real-Time Resource Slicing with Deep Dueling Neural Networks. *IEEE Journal on Selected Areas in Communications* 37(6), 1455–1470 (2019).

31. Abiko, Y., Saito, T., Ikeda, D., Ohta, K., Mizuno, T., Mineno, H.: Flexible Resource Block Allocation to Multiple Slices for Radio Access Network Slicing Using Deep Reinforcement Learning. *IEEE Access* 8, 68,183–68,198 (2020).

32. Hua, Y., Li, R., Zhao, Z., Chen, X., Zhang, H.: GAN-Powered Deep Distributional Reinforcement Learning for Resource Management in Network Slicing. *IEEE Journal on Selected Areas in Communications* 38(2), 334–349 (2020).

33. Xiang, H., Yan, S., Peng, M.: A Realization of Fog-RAN Slicing via Deep Reinforcement Learning. *IEEE Transactions on Wireless Communications* 19(4), 2515–2527 (2020).

34. Tahir, M., Habaebi, M.H., Dabbagh, M., Mughees, A., Ahad, A., Ahmed, K.I.: A Review on Application of Blockchain in 5G and Beyond Networks: Taxonomy, Field-Trials, Challenges and Opportunities. *IEEE Access* 8, 115,876–115,904 (2020).

35. Rawat, D.B., Alshaikhi, A.: Leveraging Distributed Blockchain-Based Scheme for Wireless Network Virtualization with Security and QoS Constraints. In: *2018 International Conference on Computing*. Networking and Communications (ICNC), pp. 332–336 (2018).

36. Samdanis, K., Costa-Perez, X., Sciancalepore, V.: From Network Sharing to Multi-Tenancy: The 5G Network Slice Broker. *IEEE Communications Magazine* 54(7), 32–39 (2016).

37. Nguyen, D.C., Pathirana, P.N., Ding, M., Seneviratne, A.: Blockchain for 5G and Beyond Networks: A State of the Art Survey. *Journal of Network and Computer Applications* **166**, 102,693 (2020).
38. Zanzi, L., Albanese, A., Sciancalepore, V., Costa-Pe´rez, X.: NSBchain: A Secure Blockchain Framework for Network Slicing Brokerage. In: *ICC 2020–2020. IEEE International Conference on Communications (ICC)*, pp. 1–7 (2020).
39. Gorla, P., Chamola, V., Hassija, V., Niyato, D.: Network Slicing for 5g With UE State Based Allocation and Blockchain Approach. *IEEE Network*, 1–7 (2020). DOI 10.1109/MNET.011.2000489.
40. Xu, H., Klaine, P.V., Onireti, O., Cao, B., Imran, M., Zhang, L.: Blockchain-Enabled Resource Management and Sharing for 6G Communications. *Digital Communications and Networks* **6**(3), 261–269 (2020). www.sciencedirect.com/science/article/pii/S2352864820300249.
41. Nour, B., Ksentini, A., Herbaut, N., Frangoudis, P.A., Moungla, H.: A Blockchain-Based Network Slice Broker for 5G Services. *IEEE Networking Letters* **1**(3), 99–102 (2019).
42. Adhikari, A., Rawat, D.B., Song, M.: Wireless Network Virtualization by Leveraging Blockchain Technology and Machine Learning. In: *Proceedings of the ACM Workshop on Wireless Security and Machine Learning*. Association for Computing Machinery, pp. 61–66 (2019).

Chapter 5

Key Generation

Encryption and Decryption Using Triangular Cubical Approach for Blockchain in the 6G Network and Cloud Environment

Surendra Kumar[1] and Narander Kumar[2]

CONTENTS

5.1 INTRODUCTION

This section presents the cryptosystem for the blockchain in 6G networks and cloud computing, as well as the motivation for and contribution of the key generation mechanism. The organization of this research work is also given.

The world is moving toward digitization and computerized change in the daily needs for resources. Utilizing advanced information through the web

---

[1,2] Department of Computer Science, Babasaheb Bhimrao Ambedkar University
(A Central University), Lucknow, India
Corresponding author: nk_iet@yahoo.co.in

DOI: 10.1201/9781003264392-5

or other system is becoming a part of all our personal and workaday lives. Therefore, data security turns out to be significant and progressively more significant and essential for us all when sensitive, private, and important data is conveyed through an uncertain system or stored as chronicled information. There is a need to convey electronic information or records safely and securely. From the perspective of the sender of data, guaranteeing the trustworthiness and classification of data is an essential prerequisite. For the data recipient, nonrenouncement and trustworthiness are significant features.

To achieve data security, symmetric and asymmetric encryption techniques are used. *Symmetric* key encryption uses a single key, and the one who receives messages must maintain that secret key in order to decrypt the text into the original data. *Asymmetric* key encryption involves two keys: a secret key and a private key. Digital signature asymmetric key encryption, used correctly, can provide great security to integrity and nonrepudiation. Various encryption and decryption techniques such as BlowFish, AES, DES, and 3DES are relevant for better encryption and decryption in a cloud computing environment. AES has not stopped or shown very much protection of encrypted data against hijackers and muggers in maintaining confidentiality over the transmission of data [1].

Cloud database is a highly organized data resource in the digital platform. Various techniques are used to secure data, such as encryption and digital signature, when data is transmitted over sites or accessed from them. Data security performs in tandem with various parameters and with various database management technologies to protect against unauthorized access, muggers, hijackers, and attackers [2].

Key generation algorithms are poised to play an important role in cryptography. For this concern, keys are needed to maintain various parameters such as the cycle of functions, key size, message size, numbers of cycles, etc. The cryptography algorithm does not worry about a brute force attack because its complexity is much higher than other algorithms. The main objective of attackers is to break down ciphertext and get access to plaintext. It avoids the weak keys and attempts to make strong keys weak by the various sophisticated technologies and considerable abilities of hijackers [3].

5.1.1 Main Motivation

Various techniques are used to develop secure encryption and decryption with the help of permutation, substitution, and mix structure. The key of AES is to adjust the last cycle key that can be accessed quickly. In the look-forward strategy, acquiring the all-transitional round key turns out to be more advantageous when quick decoding is required. Replacement-based key inference work is recommended to acquire a small key size as information, create key squares, and interchange the string in a key-like randomization. The key determination work is to produce irregular and flighty secret keys.

5.1.2 Main Contribution

1. For the security of the 6G network and cloud computing environment, a new mathematical model has been proposed.
2. The comprehensive security of the network is enhanced and increases the complexity of the key.
3. It is computationally very secure, and brute attack is very time-consuming, making the key very hard to crack.
4. Entropy and brute force have been tested and found that the designed mechanism is very efficient for complex key generation and that it produces a prediction-free structure.
5. The security is enhanced with the OTP request for verification for authentic access by the receiver.

This research work is organized as follows: Section I discusses the detailed introduction of preliminaries, including the overviews of objectives and the need for security in a network. Section II discusses the details of background research and problem finding. Section III shows the related works is done. Section IV presents the design of a mechanism for key generation in the cryptosystem. Practical solutions and discussions are given in Section V. Section VI presents the conclusion and future scope.

5.2 BACKGROUND OF STUDY

In this section, we discuss the major key points in cloud security. We study the different types of systems and the need for key generation to secure encrypted data from hijackers and muggers and to maintain confidentiality over data transmission.

5.2.1 Crypto System

Data security starts with the historical backdrop of PC security, which entails the need to protect hardware/software and physical areas against outside strings. To ensure centralized servers and information integrity, staggered security procedures have been developed. An association ought to have accompanying staggered security to provide security to the framework and data: physical security, individual security, task security, interchange security, system security, and data security. *Physical security* handles the security of physical things and access to authentic use or misuse. *Individual security* ensures that individuals or groups of users are approved for access. *Tasks security* and *interchange security* ensure the arrangement of exercises and correspondence media. *System security* and *data security* are utilized to ensure organizing segments and data resources. Data security is part of building a security structure for a secure network.

Cryptography is a technique used in computer science for securing data based on mathematical theory. Cryptography is a combination of three terms:

- Cryptography
- Cryptology
- Cryptanalysis

These three terms are often used interchangeably. *Cryptology* is the study of communication over nonsecure channels and related security problems. The process of designing an overall cryptosystem is called *cryptography*. Breaking the cryptosystem is *cryptanalysis*. The term "coding theory" describes cryptography. It deals with the code symbols that represent the input information symbols by means of output symbols. The coding theory explains that communication over different channels protects data from leakage and ensures that the message received is correct over nonsecure channels.

Cryptography algorithms are practically difficult to decode. In digital rights management and copyright infringement of digital media, cryptography plays a significant role. In cryptography, the key used changes the outcome of the cryptographic algorithm. Cryptography enables the secure transfer of sensitive data from cloud users to cloud providers and vice versa [4, 5].

Compared to traditional methods, newer encryption algorithms have been built more efficiently. So the latest security algorithms are all but impossible to break. Cryptography techniques convert data into a cryptic (nonreadable) form to protect original data from hackers [6]. In encryption, the cloud information is collected in ciphertext and clients are assigned the key, which is utilized to decrypt the ciphertext to plaintext to get the real message [7]. The plaintext-to-ciphertext change of information is called encryption. Decryption returns the information to its original structure. The fundamental design of the cryptography framework is shown in Figure 5.1.

- *Plaintext*: Plaintext is the original message to be sent from one end to another in cryptography. Explicit content, expressed as it is in an everyday wording, lends itself to the cryptographic algorithm process. Commonly, encryption is the process that users actuate before transmission. Plaintext is ordinary explicit content before being encoded into ciphertext or that later is decrypted.
- *Ciphertext*: Ciphertext is the encryption performed on plaintext utilizing an algorithm in cryptography, known as a cipher. Ciphertext is also called an encrypted or encoded message. It has a basic plaintext form that is

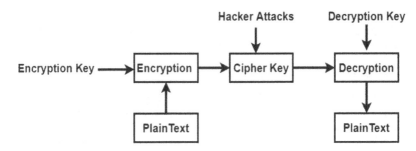

Figure 5.1 Cryptosystem framework [8].

incomprehensible to a user or system without the correct cipher to decrypt it. Decryption is the encryption inverse, i.e., the process of turning ciphertext back into understandable plaintext. The ciphertext is not to be confused with code text since the latter is the result of code result, not a cipher.

• *Key*: The key is a data piece that characterizes the practical yield of a cryptographic approach in cryptography. A key decides the change of plaintext into ciphertext by means of encryption algorithms and the other way around for decryption. Keys likewise characterize changes in different cryptographic algorithms, for example, message verification codes and digital signature.

The various terms utilized in cryptography are plaintext, ciphertext, key, encryption, and decryption. Here plaintext is a unique message that is conveyed between the sender and receiver. The ciphertext is encrypted data that is not quite the same as the original data. The encryption is done using an encryption key. The original text is obtained from the encrypted text using a decryption key. The key is a special symbol used to encrypt and decrypt the original text [8].

Hackers might use four types of attacks to hack plaintext using ciphertext:

• Chosen ciphertext
• Ciphertext only
• Chosen plaintext
• Known plaintext

Cryptography deals with encryption and decryption of messages and solving security issues that require information security. Cryptography deals with the following information security:

• Authentication of user
• Data integrity

- Data confidentiality
- Nonrepudiation of data

5.2.2 Key Generation System

Data integrity is achieved when it is uncorrupted or isolated from threat conflicts. Hacking takes place while data is being sent or stored, and authentication is the process of determining the original rather than the hacked version, since data authentication can be changed when it is transmitted across a network. Cryptography can be broadly categorized into two parts:

- Symmetric key
- Asymmetric key

Encryption techniques are widely used, and various encryption approaches like RC2, RC6, DES, Blowfish, IDEA, and AES are based on symmetric keys. Symmetric key cryptography has used these keys to perform encryption and decryption algorithms to convert data from plain text to ciphertext form. Asymmetric key cryptography uses two keys to accomplish these processes. RSA is an important asymmetric encryption algorithm that uses a couple of keys, i.e., public key and secret key [9].

Encryption uses a public key in plaintext conversion and decryption using a private key for ciphertext conversion. Both public and private keys are based on the concept of prime numbers and are calculated by multiplying the prime numbers. The size of the key is 1024 to 2048 bits. AES (Advanced Encryption Standard) uses an advanced encryption scheme: 128, 192, or 256 variable key bits, which are used to encrypt data blocks with 128,192 or 256 symmetric keys. The encryption is done based on the key size in 10, 12, and 14 rounds. The DES (Data Encryption Standard) encryption algorithm has 56-bit keys. The triple-DES is an advancement of the DES encryption algorithm [10]. The key length of DES is increased by using the algorithm thrice with three keys. Here the encryption algorithm is carried out at three levels, and the DES key size is extended in the Triple-DES algorithm using the DES algorithm thrice with three keys.

RC2 was introduced in the year 1987 by Rivest. The full form of RC is "Ron's code," which is a symmetric key algorithm as well as 64-bit block cipher. Rivest also designs RC6, with a block size of 128 bits and a structure the same as that of RC2. In RC6, multiplication operation is carried out for rotation operation, which is not included in RC2. Bruce Schneier introduced the blowfish algorithm in 1993, which uses a symmetric key block cipher in nature and provides an efficient encryption scheme [11]. The size of the block cipher is 64 bit, and key length ranges from 32 to 448.

Cryptography is divided into three categories. Conversion methods, whether key based and process based, convert plaintext into cipher in three

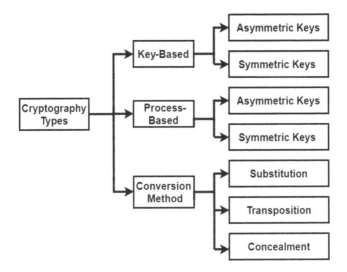

Figure 5.2 Classification of cryptosystem [12].

ways: substitution, transposition, and concealment. In the key-based method, the symmetric and asymmetric keys are used to a hide the message using the encryption and decryption process. In the processed-based method, plain text is processed by block and stream ciphers, as given in Figure 5.2.

Encryption and decryption are performed by various symmetric key and asymmetric key algorithms. It is easier to crack keys that do not present troublesome multiple possibilities [12].

5.2.3 Message Digest Functions

Message digests create a digital summary report of a message, known as hash functions. Message digests range lies between 128 and 160 bits long and gives an advance identifier for each record or digital document [13]. A message digest is an arithmetical operation that can create advanced data to make a different message digest for each record. Similar types of data contain the indistinguishable message digest; document modifications contain some of the bits.

5.2.4 Secret Key Exchange Algorithms

For symmetric key cryptography to execute network transmissions, the secret key must be immutably imparted to the approved conveying authority and must be ensured against discovery and utilization by an unapproved authority. Public key cryptography is employed to give a protected technique for exchanging secret keys online. The two basic key exchange algorithms are as follows:

1. RSA Key Exchange Process
2. Diffie–Hellman Key Agreement Algorithm

Both techniques offer a highly secure key exchange between communicating parties. An intruder who interrupts network transmission cannot simply estimate or decrypt the secret key necessary to decrypt transmissions. The precise mechanisms and algorithms used for key exchange vary for every security methodology. The Diffie–Hellman Key Agreement Algorithm provides advancement in various aspects of performance from the RSA key exchange algorithm [14].

5.2.5 HMAC-Hashed Message Authentication Code Functions

HMAC makes a message digest for all sets of sent information and uses a subjective private symmetric key to encode the message digests over the transmission process. This private key is immutably shared among the gatherings contained in the secured transmission. While information is acquired, the private key is fundamental to decrypt the message process and execute the information trustworthiness confirmation. HMAC cryptographic quality depends on the hidden quality of the message digest used and how immutably the private key is exchanged [15]. A gatecrasher doesn't perceive the private key and can't meddle with the information in transit or reproduce the message digest. HMAC gives information respectability and security in a way that is practically identical to that of a computerized signature, yet it doesn't require conveying gatherings to contain private and open keys. HMACs additionally provide an improved introduction for mass online correspondences, compared to open key computerized marking strategies.

Users can create a primary HMAC, or hash function per request, for a hashing data request using the secret key and send the acknowledgment of request. This feature provides more security than the message authentication code utilized by the key, and messages are hashed in individual steps.

$$HMAC\ (k, k\_msg) = H(mod1(k) \parallel H(mod2(k) \parallel k\_msg))$$

5.2.6 Digital Signatures

Digital signatures further enable procedures to create advanced signatures. Today's various basic digital signatures are framed by marking message digests with the private key of the beginning of production of a digital information thumbprint. Since just the message digest is signed, the digital signature is normally a lot shorter than the agreed-upon information. In addition, digital signatures set a similarly low burden on PC processors during the signing procedure, go through insignificant measures of transfer speed, and make modest quantities of ciphertext for the cryptanalysis process [16]. Today, some widely used advanced digital signature approaches are the Digital Signature Algorithm (DSA) and the RSA digital signature.

5.2.7 ECC: Elliptic Curve Cryptography

ECC creates keys using the parameter of the elliptic curve situation as an option of the typical approach for creation as the outcome of incredibly enormous prime numbers in the form of key attributes. The approaches can be used in connection with enormous loads of open key encryption procedures, like Diffie–Hellman and RSA. ECC can check security levels with a 164-bit key, whereas various other frameworks need a 1024 bit key for such a purpose [17].

ECC needs fewer significant keys than non-ECC cryptography to afford the same security. Elliptic curves are appropriate for encoding, digital marks, and other tasks. They are also used in some integer factorization techniques that have relevance in cryptography. Public key cryptography is based on tortuous mathematics. Early public key systems are protected, assuming that it is hard to factor a big integer composed of two or more huge prime factors.

5.3 RELATED WORK

This section presents an extensive review of work done in key generation for encryption and decryption. The findings are that blockchain in 6G network and cloud computing require more attention to apply the key generation technique in encryption and decryption.

Key generation is used in cryptosystems. The key contains various combinations and plain text in the form of words, numbers, and combinations. These keys convert the plaintext or message into ciphertext based on their system strength and key isolation [18]. A linear encrypted file sharing multiple key values improves access in cloud computing and its categories in attribute name and values [19].

Design encryption is based on the attributes of outsourced decryption to satisfy the properties and is reduced to basic attribute-based encryption in the black box security model. The decryption mode has been updated to satisfy adaptive security [20]. Improving the reusability of resources and using a secure mechanism for the minimum threshold for various clients fulfill the requirements policy for decryption to ciphertext by minimizing the decryption cost for users. It can also analyze an array of servers for requesting users who fulfill access agreements and request decryption services [21].

Decryption designed to be verifiable for attribute-based encryption checks the accuracy of changes in ciphertext from authorized or unauthorized access in cloud computing. It shows more of the secured ciphertext attributes in the basic model [22]. The hardware and software code design framework were designed to optimize encryption and decryption operation with high-performance polynomial multipliers. One-byte polynomial with an 8 bit coefficient for plain text and a 32-bit coefficient for ciphertext, as well as I/O operations for offload encryption and decryption, are performed [23].

Untrusted model and transformation keys are designed for molecular exponentiation with verifiable encryption/decryption to service providers performing only restricted numbers of service requests for authentic users. The user and servers are assured correctness and completeness of results from encryption and decryption service providers [24]. Binary code and arbitrary keys are used to generate the ciphertext. A key is used to establish a platform between users to secure the data from hijackers and muggers as they maintain confidentiality over data transmission [25].

The re-encryption method is used without download data by attribute-based encryption scheme to reduced transmission cost between user and servers [26]. Srinivas et al. discussed a new technique using the elliptic curve for encryption and decryption that increases speed and minimizes memory requirements and that maintains security with the help of a public key cryptosystem [27].

Encryption and decryption are performed by using ASCII, binary bit sequence, and XOR operations. Plain text is converted into ciphertext with encryption methods by using random keys, the same random keys used to decrypt text to plain text. This process is known as an asymmetric-key algorithm [28]. Introduced hybrid homomorphic encryption using Goldwasser–Micali (GM) encryption is a type of single-bit homomorphic encryption. It increases limitation on and resistance against attackers and can also improve the speed and execution time and provide two-layer encryption security [29].

The hybrid key generation and key verification approach shows that secure information in verification is negligible and that the verification key is unique for the Internet of Industrial Things [30]. Key generation and problem analysis are worked simultaneously. Public and private keys are generated to bridge legitimate terminals so that the trusted user can be enabled to access the data completely in the public discussion. An authentic user can seek to generate a public key hidden from the end user, and source and destination ends are needed to create a private key that affects well enough alone from the third real ends [31].

Two-way randomness techniques are used in secret key generation for single input and a single output. The real handsets commonly transmit their irregular sign through a proportional remote channel. Afterward, the augmentation of sending and receiving networks is utilized for basic changeability to secure keys [32]. IoT faces more malicious attacks than the encryption required for communication and the keys needed to manage the encryption can handle, so they generate a key generation process and manage it with the help of block-chain [33].

Various encryption and decryption techniques are analyzed in different contexts, such as speed key computational time, memory, etc. These need to be considered, along with the effect on the efficiency of an algorithm and the assurance of optimal allocation. It is very difficult for malicious attackers to understand ciphertext created using the symmetric key and asymmetric key and single and two discrete keys [34]. The cubical surface is separated into four parts by intersecting the primary and secondary diagonals, resulting in a very complex key generation process and enhancing the security of the ciphertext in the encryption process as the key size is increased, so that time complexity is increased in cloud computing [35]. Table 5.1 illustrates the comparison of some of the existing methods.

Table 5.1 Comparison of the Existing Encryption and Decryption Methods Using Key

SI No	Authors	Proposed techniques	Advantages	Limitations
1	Patnala et al. [18]	To highly secure data transmission in ASIC design flow modernistic paths for key generation	Improved security of the system; discovering the key is hard.	Data can be predicted easily by the attackers.
2	Zhang et al. [19]	Personal health record system a hidden CP-ABE with fast decryption	Using static assumptions attains full security in the standard model.	The proposed technique doesn't accomplish completely the hiding policy with quick encryption.
3	Baodong Qin et al. [20]	The decryption of ABE using generic technique to outsource	The selective security model solves the practical issue.	In this approach, delivering adaptive security for normal outsourced decryption mode is difficult.
4	Yongjian Liao et al. [21]	ABE schemes for a user and cloud server to manage cost-efficient outsourced decryption in green cloud computing	The approach minimizes the overhead of client and data centers.	In the green computing networks for the server, this technique is not effective.

(Continued)

Table 5.1 (Continued) Comparison of the Existing Encryption and Decryption Methods Using Key

Sl No	Authors	Proposed techniques	Advantages	Limitations
5	Jiguo Li et al. [22]	ABE to full verifiability for outsourced decryption	This methodology check the rightness for changed ciphertext for the approved clients and unapproved clients.	The proposed approach needs to focus in addition on the rightness for the changed ciphertext.
6	Mert et al. [23]	Encryption/ decryption architectures is design for BFV homomorphic encryption scheme	For the encryption and decryption tasks, quick and profoundly parallelized hardware framework is utilized.	Homomorphic multiplication should be accelerated for high performance.
7	Zhidan et al. [24]	Verifiable outsourced for encryption and decryption in an efficient ABE scheme	Delete the most overhead execution on both the information owner and the client sides.	Transformation key generation should be considered for computational cost.
8	Mohan Kumar et al. [25]	DNA technology for file encryption and decryption	The designed approach is all the more effective and productive since it can store more information.	Security for storing huge amounts of data should be improved.
9	Yasumura et al. [26]	Revocation with attribute-based proxy re-encryption technique in cloud data center	Reduces the transmission cost by one-quarter in comparison with the trivial solution.	More computation time is needed.
10	Srinivas et al. [27]	Using elliptic curves for encryption and decryption in public key cryptosystems	Minimize the memory needs for execution and improved encryption and decryption speed.	Increase the memory size for computation, and minimize the speed of encryption and decryption.
11	Hussain et al. [28]	Binary bit sequence and multistage encryption used for IoT network designed a new cryptosystem	Reduce consumption of memory with high throughput.	Requires extensive computation.

(Continued)

Table 5.1 (Continued) Comparison of the Existing Encryption and Decryption Methods Using Key

Sl No	Authors	Proposed techniques	Advantages	Limitations
12	Mahmood et al. [29]	Multistage partial homomorphic encryption with novel homomorphic encryption scheme based used in cloud computing	Enhance the speed times, minimize the computation time, enhance confidentiality of the data.	Not suitable for a digital computing environment.
13	Kurt et al. [30]	Hybridization of a verification scheme and key generation scheme	Uncovered data during the key check measure and ensured indistinguishable keys.	Resource utilization is higher than the data compromise schemes.
14	Gong et al. [31]	A trusted model in source type model for maintaining secure secret key and private key generation	Improves the security efficiency of the generated keys as far as the key leakage rate.	The source model is needed for multikey generation.
15	Zhang et al. [32]	TDD-SISO mechanism to two-way randomness approach for secure secret key generation	The proposed approach produces higher secret key.	Multi-input multi-output scenarios are missing.
16	Choil et al. [33]	Blockchain to random seed generation for key generation and management system in IoT	The designed mechanism diminishes the risk of information spillage and produces irregular qualities.	Computationally intensive and less efficient.
18	Mushtaq et al. [35]	His technique on 2D triangular extraction for key generation in a cryptosystem	Security is improved and upgrades the intricacy in the plan of the key scheduling approach.	Requires high computation cost.
19	Proposed approach	Key generation: encryption and decryption using triangular cubical approach for blockchain in 6G network and cloud environment	The overall security of the proposed method is improved and increases complexity of the key.	

5.4 PROPOSED KEY GENERATION METHOD

This section shows the proposed key generation method called the cubical matrix approach and gives the step-by-step process of key generation using the cubical matrix method.

The proposed cubical matrix approach is categorized in different stages (P1, P2, P3, P4) and in the diagonally bisect matrix, which goes to the center. These intersections are lying on the X and Y pivot and employ an applied proposed mechanism on staying structural surfaces. Cubical key rotation for a triangular framework is right circular, and the principle key values are used to generate encryption keys. See Figures 5.3–5.7. Various stages of key generations are shown in Table 5.2.

In key generation matrix A = $[a_{mn}]_{rc}$ here r = row, c= column if m = n than matrix key is (1,1)(2,2)(3,3)(4,4) as well as other stages if selecting keys is $a_{m,c-m+1}$ and m $\in$ {1,2, ... ,n} key is (1,4) (2,3) (3,2) (3,4).

Step 1: Let's take a set of 4×4 key matrix for key generation for calculating the symmetric keys matrix (m,n) and (n,m) selecting key: {(1,2)(2,1)}, {(1,3)(3,1)},

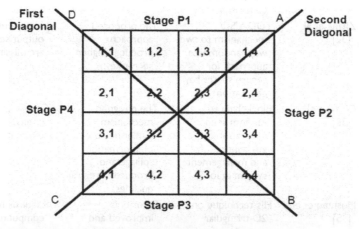

Figure 5.3 Proposed matrix key generation process for encryption and decryption.

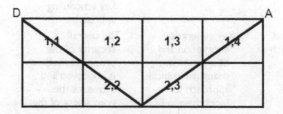

Figure 5.4 Stage PI matrix key triangular selection for D to A.

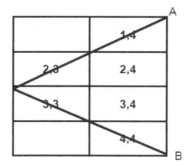

Figure 5.5 Stage P2 matrix key triangular selection for A to B.

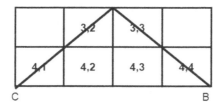

Figure 5.6 Stage P3 matrix key triangular selection for B to C.

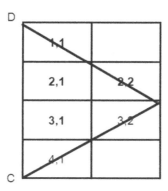

Figure 5.7 Stage P4 matrix key triangular selection for C to D.

{(1,4)(4,1)}, {(2,3)(3,2)}, {(3,4)(4,3)}, {(4,2)(2,4)}. So we are taking the mean value of every symmetric two times for secure key generation matrix.

First mean of symmetric key: [{3}{4}{5}{5}{7}{6}]

Second mean of symmetric key: $\dfrac{3+4+5+5+7+6}{6} = 5$

Step 2: For the key generation set in a key generation (m,n) where m = n is taken from the first diagonal, which is a collection of cubical entities, as well as if a key is m + n = 5 from the second diagonal, which is a surface matrix by applying step 1.

If key matrix is m = n and m + n = 5, then the coordinate of the particular key cell set is $\frac{1}{2}$ (m,n) so that intersecting lines fulfill all symmetric and reflective definition: {(1,1) (2,2) (3,3) (4,4) (3,2) (2,3) (4,1) (1,4)}.

5.5 RESULTS AND DISCUSSION

This section discusses the outcomes of the proposed mechanism and generates the key, which is more complex and more difficult to crack for unauthorized access or theft. Entropy is also calculated for better results.

The technique is proposed for key generation using various matrixes for encryption algorithms as well as for decryption. The encryption and decryption key uses the matrix for the nonbinary block cipher. Let's take an example [36] and look at the step-by-step process to figure it out.

$$A = \begin{bmatrix} 410 & 1364 & 3150 & 437 \\ 420 & 3249 & 2304 & 200 \\ 1890 & 64 & 289 & 3660 \\ 2805 & 594 & 300 & 1932 \end{bmatrix}$$

$$B = \begin{bmatrix} 822 & 3126 & 1079 & 146 \\ 201 & 1364 & 3140 & 790 \\ 3586 & 594 & 300 & 1635 \\ 1322 & 81 & 1024 & 2466 \end{bmatrix}$$

$$C = \begin{bmatrix} 1635 & 66 & 548 & 3586 \\ 3650 & 586 & 1 & 1990 \\ 200 & 1691 & 4086 & 410 \\ 790 & 3196 & 1578 & 201 \end{bmatrix}$$

$$D = \begin{bmatrix} 2746 & 635 & 246 & 1342 \\ 1942 & 76 & 548 & 2715 \\ 447 & 3156 & 1578 & 130 \\ 176 & 1650 & 2421 & 832 \end{bmatrix}$$

Table 5.2 Stages (P1,P2,P3,P4) Selection and Key Generation Formulations

Stages	Selection points of the matrix
P1	$\displaystyle\sum_{m=0}^{1}\sum_{n=1+m}^{4-m}(m+1,n)$ ... Eq. (5.1)
P2	$\displaystyle\sum_{n=0}^{1}\sum_{m=1+n}^{4-n}(m,n+1)$ ... Eq. (5.2)
P3	$\displaystyle\sum_{m=0}^{1}\sum_{n=1+m}^{4-m}(4-m,n)$ ... Eq. (5.3)
P4	$\displaystyle\sum_{n=0}^{1}\sum_{m=1+n}^{4-n}(m,n+1)$ ... Eq. (5.4)

Step 1: Four matrixes, A, B, C, D, are generated for the encryption algorithm, and every matrix is solved by P1, P2, P3, P4 with the formulation of Eqs. (5.1), (5.2), (5.3), and (5.4) and getting a rotated matrix.

$$\text{Rotated matrix} = \begin{bmatrix} 7833 & 5497 & 3073 & 6017 \\ 7037 & 5771 & 3349 & 5063 \\ 3249 & 5881 & 7945 & 6193 \\ 3173 & 5355 & 6733 & 4583 \end{bmatrix}$$

Step 2: Take modulo 16 of the rotated matrix, and check the repetition of the generated keys matrix.

$$\text{Repeated key} = \begin{bmatrix} 9 & 9 & 1 & 1 \\ 13 & 11 & 5 & 7 \\ 1 & 9 & 9 & 1 \\ 5 & 11 & 13 & 7 \end{bmatrix}$$

Step 3: Remove the repeated key using the repetition value = repeated value – number of repetitions.

$$\text{Unrepeated key} = \begin{bmatrix} 9 & 8 & 1 & 0 \\ 13 & 11 & 5 & 7 \\ 2 & 6 & 4 & 3 \\ 10 & 12 & 14 & 15 \end{bmatrix}$$

Step 4: Select the triangular quarter's keys from Table 5.2, and calculate with the rotation matrix key and the unrepeated key.

$$
\text{Key matrix=}
\begin{bmatrix}
8391 & 1618 & 4899 & 3080 \\
10293 & 2547 & 10002 & 1618 \\
8430 & 6333 & 3924 & 3924 \\
10913 & 2547 & 6921 & 3080
\end{bmatrix}
$$

Step 5: Rotate the key matrix in a clockwise direction, and get a unique key matrix.

$$
\text{Final key matrix=}
\begin{bmatrix}
10913 & 8430 & 10293 & 8391 \\
2547 & 6333 & 2547 & 1618 \\
6921 & 3924 & 10002 & 4899 \\
3080 & 3924 & 1618 & 3080
\end{bmatrix}
$$

This process produces very high complexities, which provide greater security features and increase the ciphertext efficiency. Integer numbers used in encryption keys for matrix i,j and keys lie between 212 bits, and the keyspace is approximately 2^{192} or 10^{59} keys. The entropy of the random matrix results is made very efficient in the encryption and decryption process by using the calculate() function in MatLab. The entropy of the final key matrix is 0.9512, which is nearer to 1, not to zero. So, this block size of 16 decimal numbers is 95.12% random. Figure 5.8 shows the encrytion and decryption using proposed key generation mechanism and Figure 5.9 shows the comparison of the encryption and decryption time using the complex security key generated with

```
dcs@DCS:/media/dcs/B6327A2D3279F32B/security 1/modified AESS python aes1.py
OTP Program.  Key must be less than or equal to plaintext in length.  Plaintext must not contain numbers.

Choices:
1: Encrypt
2: Decrypt
3: Quit
>>> 1
Please enter your plaintext: Welcome to BBAU, Lucknow
Enter your OTP : 564
Dh3kzbNh8Kp5BeWFBoNHOtSeoAmJfVV5sXQ8A8YTpQwbIKuJJXWUAAAl1q79BVor
>>> 2
Please enter your encrypted Message: Dh3kzbNh8Kp5BeWFBoNHOtSeoAmJfVV5sXQ8A8YTpQwbIKuJJXWUAAAl1q79BVor
Enter your OTP : 564
Welcome to BBAU, Lucknow
```

Figure 5.8 Python code output for encryption and decryption using key generation.

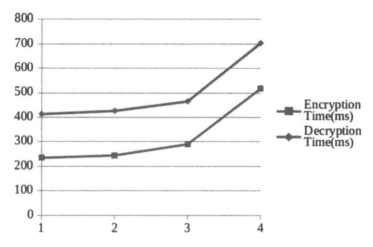

Figure 5.9 Analysis of encryption and decryption times using key generation.

the proposed mechanism, which shows optimized time and provides efficient security to the original message.

5.6 CONCLUSIONS AND FUTURE PERSPECTIVES

The cloud computing and 6G network data storage security models and challenges have been studied. With today's advanced software technology, security algorithms are easily broken by the attacker. A new mathematical model has been proposed to upgrade the security of the 6G network and cloud computing environment. The proposed mechanism for encryption is analyzed more securely and increases key complexity for data theft. This chapter proposes an approach based on four triangular stages (P1, P2, P3, P4) made up of two diagonals that intersect in the matrix. These four parts are used to select rotation in the cubical key matrix, and these rotations secure the message from the attacks during transmission. The encryption and decryption calculated keys space are approximately $2^{12} \times 2^{12} \times 2^{12} \times 2^{12} \ldots \times 2^{12} = (2^{12})^{16}$ or approximately $10^{512} \times 10^{512} \times 10^{512} = (10^3)^{512}$ or 10^{1536}, so it is computationally very secure, and brute attack is made very time-consuming and the encryption very hard to crack. We also generate an OTP for more secure encryption and decryption processes in a cloud computing environment. The OTP platform key must be less than or the same length as plaintext, and the plaintext must not contain numbers. In future, the preceding work may be considered to improve a property, apply the modified encryption cipher round and key generation algorithm in secret files, and determine the added operation in terms of performance to throughput.

REFERENCES

1. Then, A. A., Thwin, M. M. S.: Modification of AES Algorithm by Using the Second Key and Modified Sub Bytes Operation for Text Encryption. In *Computational Science and Technology*, pp. 435–444. Springer (2019).
2. Zaw, T. M., Thant, M., Bezzateev, S. V.: Database Security with AES Encryption, Elliptic Curve Encryption and Signature. In *2019 Wave Electronics and Its Application in Information and Telecommunication Systems (WECONF)*, pp. 1–6. IEEE (2019).
3. Saha, R., Geetha, G., Kumar, G., Kim, T. H.: RK-AES: An Improved Version of AES Using a New Key Generation Process with Random Keys. *Security and Communication Networks*, 1–11 (2018).
4. Sugumar, R., Sheik Imam, S. B.: Symmetric Encryption Algorithm to Secure Outsourced Data in Public Cloud Storage. *Indian Journal of Science and Technology*, 8(23), 1–5 (2015).
5. European Bioinformatics Institute. www.ebi.ac.uk/genomes/: Accessed on 15/01/2016.
6. Lim, H. W., Paterson, K. G.: Identity-Based Cryptography for Grid Security. In *Proceedings of the First International Conference on e-Science and Grid Computing (ESCIENCE)*, vol. 10, no. 1, pp. 395–404. IEEE Press (2005).
7. Kumar, L., Rishiwal, V.: Design of Retrievable Data Perturbation Approach and TPA for Public Cloud Data Security. *Wireless Personal Communications*, 108(1), 235–251, (2019).
8. Sun, Y., Zhang J., Xiong, Y.: Data Security in Cloud Using RSA Computing. *Proceedings of the International Conference on Communications and Networking Technologies (ICCCNT)*, 10(7), 1–9 (2013).
9. Shimanovsky, B., Feng, J., Potkonjak, M.: Hiding Data in DNA. Proceedings of the 5th International Workshop on Information Hiding, 2578, pp. 373–386, (2002).
10. Yassein, M. B., Aljawarneh, S., Qawasmeh, E., Mardini, W., Khamayseh, Y.: Comprehensive Study of Symmetric Key and Asymmetric Key Encryption Algorithms. In *2017 International Conference on Engineering and Technology (ICET)*, pp. 1–7. IEEE (2017).
11. Knudsen, L. R., Rijmen, V., Rivest, R. L., Robshaw, M. J.: On the Design and Security of RC2. In *International Workshop on Fast Software Encryption*, pp. 206–221. Springer (1998).
12. Venkatesh, A. Eastaff M. S.: A Study of Data Storage Security Issues in Cloud Computing. *International Journal of Scientific Research in Computer Science, Engineering and Information Technology*, 3(1), 1741–1745 (2018).
13. Rupa, C., Avadhani, P. S.: Message Encryption Scheme Using Cheating Text. In *2009 Sixth International Conference on Information Technology: New Generations*, pp. 470–474. IEEE (2009).
14. Rawat, A. S., Deshmukh, M.: Efficient Extended Diffie-Hellman Key Exchange Protocol. In *2019 International Conference on Computing, Power and Communication Technologies (GUCON)*, pp. 447–451. IEEE (2019).
15. Latif, M. K., Jacinto, H. S., Daoud, L., Rafla, N.: Optimization of a Quantum-Secure Sponge-Based Hash Message Authentication Protocol. In *2018 IEEE 61st*

International Midwest Symposium on Circuits and Systems (MWSCAS), pp. 984–987. IEEE (2018).

16. Ruballo, Á. H. Z.: Application for Digital Signature and Generation of Certificates Using the Bouncy Castle API Considering Digital Signature Law in El Salvador. In *2018 IEEE 38th Central America and Panama Convention (CONCAPAN XXXVIII)*, pp. 1–6. IEEE (2018).

17. Eldeen, M. A., Elkouny, A. A., Elramly, S.: DES Algorithm Security Fortification Using Elliptic Curve Cryptography. In *2015 Tenth International Conference on Computer Engineering & Systems (ICCES)*, pp. 335–340. IEEE (2015).

18. Patnala, T. R., Jayanthi, D., Majji, S., Valleti, M., Kothapalli, S., Karanam, S. R.: A Modernistic Way for KEY Generation for Highly Secure Data Transfer in ASIC Design Flow. In *2020 6th International Conference on Advanced Computing and Communication Systems (ICACCS)*, pp. 892–897. IEEE (2020).

19. Zhang, L., Hu, G., Mu, Y., Rezaeibagha, F.: Hidden Ciphertext Policy Attribute-Based Encryption with Fast Decryption for Personal Health Record System. *IEEE Access*, 7, 33202–33213 (2019).

20. Qin, B., Zheng, D.: Generic Approach to Outsource the Decryption of Attribute-Based Encryption in Cloud Computing. *IEEE Access*, 7, 42331–42342 (2019).

21. Liao, Y., Zhang, G., Chen, H.: Cost-Efficient Outsourced Decryption of Attribute-Based Encryption Schemes for Both Users and Cloud Server in Green Cloud Computing. *IEEE Access*, 8, 20862–20869 (2020).

22. Li, J., Wang, Y., Zhang, Y., Han, J.: Full Verifiability for Outsourced Decryption in Attribute Based Encryption. *IEEE Transactions on Services Computing*, 13(3), 478–487 (2020).

23. Mert, A. C., Öztürk, E., Savaş, E.: Design and Implementation of Encryption/Decryption Architectures for BFV Homomorphic Encryption Scheme. *IEEE Transactions on Very Large Scale Integration (VLSI) Systems*, 28(2), 353–362 (2019).

24. Li, Z., Li, W., Jin, Z., Zhang, H., Wen, Q.: An Efficient ABE Scheme with Verifiable Outsourced Encryption and Decryption. *IEEE Access*, 7, 29023–29037 (2019).

25. Kumar, B. M., Sri, B. R. S., Katamaraju, G. M. S. A., Rani, P., Harinadh, N., Saibabu, C.: File Encryption and Decryption Using DNA Technology. In *2020 2nd International Conference on Innovative Mechanisms for Industry Applications (ICIMIA)*, pp. 382–385. IEEE (2020).

26. Yasumura, Y., Imabayashi, H., Yamana, H.: Attribute-Based Proxy Re-Encryption Method for Revocation in Cloud Data Storage. In *2017 IEEE International Conference on Big Data*, pp. 4858–4860. IEEE (2017).

27. Srinivas, M., Porika, S.: Encryption and Decryption Using Elliptic Curves for Public Key Cryptosystems. In *2017 International Conference on Intelligent Computing and Control Systems (ICICCS)*, pp. 1300–1303. IEEE (2017).

28. Hussain, I., Negi, M. C., Pandey, N.: Proposing an Encryption/Decryption Scheme for IoT Communications Using Binary-Bit Sequence and Multistage Encryption. In *2018 7th International Conference on Reliability, Infocom Technologies and Optimization (Trends and Future Directions)(ICRITO)*, pp. 709–713. IEEE (2018).

29. Mahmood, Z. H., Ibrahem, M. K.: New Fully Homomorphic Encryption Scheme Based on Multistage Partial Homomorphic Encryption Applied in Cloud

Computing. In *2018 1st Annual International Conference on Information and Sciences (AiCIS)*, pp. 182–186. IEEE (2018).

30. Kurt, G. K., Khosroshahi, Y., Ozdemir, E., Tavakkoli, N., Topal, O. A.: A Hybrid Key Generation and a Verification Scheme. *IEEE Transactions on Industrial Informatics*, 16(1), 703–714 (2019).

31. Gong, S., Tao, X., Li, N., Wang, H., Han, Z.: Secure Secret Key and Private Key Generation in Source-Type Model with a Trusted Helper. *IEEE Access*, 8, 34611–34628 (2020).

32. Zhang, S., Jin, L., Lou, Y., Zhong, Z.: Secret Key Generation Based on Two-Way Randomness for TDD-SISO System. *China Communications*, 15(7), 202–216 (2018).

33. Choi, J., Shin, W., Kim, J., Kim, K. H.: Random Seed Generation for IoT Key Generation and Key Management System Using Blockchain. In *2020 International Conference on Information Networking (ICOIN)*, pp. 663–665. IEEE (2020).

34. Al-Shabi, M. A.: A Survey on Symmetric and Asymmetric Cryptography Algorithms in Information Security. *International Journal of Scientific and Research Publications*, 9(3), 576–589 (2019).

35. Mushtaq, M. F., Jamel, S., Mohamad, K. M., Khalid, S. K. A., Deris, M. M.: Key Generation Technique Based on Triangular Coordinate Extraction for Hybrid Cubes. *Journal of Telecommunication, Electronic and Computer Engineering (JTEC)*, 9(3–4), 195–200 (2017).

36. Jamel, S., Deris, M. M., Tri, I., Yanto, R., Herawan, T.: HiSea: A Non Binary Toy Cipher. *Journal of Computing*, 3(6), 20–27 (2011).

Chapter 6

Blockchain-Based Efficient Resource Allocation and Minimum Energy Utilization in Cloud Computing Using SMO

Narander Kumar[1] and Jitendra Kumar Samriya[2]

CONTENTS

[1,2] Department of Computer Science, Babasaheb Bhimrao Ambedkar University (A Central University), Lucknow, India
Corresponding author: nk_iet@yahoo.co.in

DOI: 10.1201/9781003264392-6

6.1 INTRODUCTION

Cloud computing in recent years has grown as a new concept for offering services over the Internet, using a dynamic pool of virtualized computational resources. The growing demands of the customers are satisfied by cloud service providers through rapidly enlarging their data centers, which leads to a high intake of energy besides releases of an enormous volume of CO_2. The latest report on this reveals that, globally, the energy expended by the big data centers is equal to the depletion of 25,000 habitations [1]. The 6G technology comes with a new revolution to provide an intelligent wireless platform for users with enhanced bandwidth and capabilities using AI applications [2].

These applications, without any software installation, can be used as a part of this cloud domain. Messages can be sent to all parts of the world by users using the Internet [3]. Google provides scalable and efficient answers for working with extensive scale data [4]. Some of the permitted effectual computing in the cloud are centralized storage, processing, memory, and bandwidth. Virtualization is considered their hidden innovation, which has driven the cloud to some degree [5]. In this domain, efficient scheduling of resources and balanced data scheduling are the main factors of resource scheduling and transmission based on feature extraction and adaptive clustering of big data. Using big data information processing and resource classification models, sensible scheduling can be attained. This enlightens the resource management and the information processing abilities of computing resources in the cloud environment.

Nowadays, the resource allocation approaches are based on a decision tree model, dynamic sliding window, support vector machine, and so on [6, 9]. The resource allocation process uses the network's computing resources to execute complex tasks [10]. Some of the factors considered for resource allocation are Makespan, load balancing, and energy consumption. In a cloud, the main factor to consider is picking the resource nodes favorable to execute a task and based on the properties of the task. They have to be selected properly [11]. The allocation of cloud resources is done to satisfy the user-specified (QoS) requirements via service-level agreements (SLAs) and also to minimize energy depletion [12, 13].

The broker, which is signified as an element of the cloud infrastructures, is liable for mapping the end user tasks to available virtualized hardware and is applied normally in the form of virtual machines (VMs). The broker does mapping by implementing the scheduling algorithm. With the development of several available resources and submitted tasks, the execution of mapping becomes complex. Some VMs may be overutilized or underutilized when an unsuitable scheduling algorithm threatens to ruin the whole system. For resource scheduling, many algorithms and techniques are available in cloud computing environments.

Classic (deterministic) algorithms are employed in some systems. They are not effective and capable of yielding substantial, optimal, or near-optimal solutions within an appropriate computational time for NP-hard challenges. Instead of using classical optimization techniques for resource scheduling in the cloud when tackling NP hard tasks, smart mechanisms that evaluate only favorable parts of the search space (not the whole search domain) could be utilized. Resource scheduling problems can be tackled only by implementing heuristics- and metaheuristics-based approaches that could produce satisfying solutions within the polynomial time. According to the literature survey, metaheuristics are bio-inspired and, inspired by nature, have been confirmed to be robust methods that can capably deal with the resource scheduling issues of the cloud.

6.1.1 Motivation

When compared to traditional cloud computing infrastructure, the blockchain-based cloud is thin. In the year 2008, blockchain was developed mainly for Bitcoin cryptocurrency to facilitate the payment system, which allows unacquainted persons to form a stable, transparent, and constant exchange record and to manage it without any central authority. There are certain layers in the blockchain system, such as the network layer, data layer, incentive layer, consensus layer, application layer, and contract layer [14, 15]. The data layer constructs the data blocks. Distributed peer-to-peer networks are applied in the network layer for communication and data verification among the nodes [16]. The incentive layer develops the incentive approach. The contract layer uses scripts or algorithms to formulate smart contracts. Based on the blockchain technology's application scenarios, the application layer is constructed. Relevant studies have been carried out with the outline of blockchain technology with cloud computing, edge computing, and fog computing. It also includes studies combining access control technology, the Internet of Things (IoT), and other relevant fields [17].

6.1.2 Contribution

The study's contribution is to achieve a blockchain and 6G-based resource management framework and an optimized resource allocation strategy using a spider monkey optimatization (SMO) algorithm based on energy consumption with a makespan optimization model. The SMO is a novel evolutionary algorithm based on spider monkey's foraging behavior. It is a perfect approach for the optimization of benchmark functions and antenna design problems. The use of SMO in this study successfully optimizes resource allocation when evaluated with the prevailing resource allocation algorithms. In addition, the energy depletion of resources is minimized by applying a brownout-based energy model. The simulation outcomes indicate that the proposed approach has enormous ability. It bodes high capability to enhance energy efficiency and to make substantial cost savings, which can satisfy the customers' SLA requests.

This chapter is organized as follows. In Section 2, related works are analyzed. Section 3 describes the background work of this study. In Section 4, resource management using blockchain is presented along with the Brownout energy model. Section 5 presents the results and discussion and Section 6 the conclusion and future work.

6.2 RELATED WORK

A brief review is given in this section about the existing resource allocation approaches using the blockchain and other new methods with energy-efficient resources allocation strategies in cloud computing. The researchers have proposed many solutions to overcome the issues of resource allocation and scheduling, but some further improvements can be made.

The authors conducted a study in [18] on a blockchain-based cloud manufacturing architecture. This study intended to augment decentralization and information transparency. The proposed platform identifies the manufacturing resources, and they are shrouded as services. Any user can purchase the manufacturing services by accessing the platform. To record transaction results and intermediate the service composition, blockchain is used in this study.

A decentralized resource management agenda using blockchain has been developed [19]. The main issues in cloud data centers (DCs) have to do with cost minimization. This study minimizes the energy consumption cost of the traditional power grid, requests migration, and reduces scheduling cost in DCs. This framework presented a migration method dependent on reinforcement learning (RL) and embedded with a smart contra for saving the contact. In favor of real-world electricity prices and Google cluster traces, the simulation is conducted. Finally, the experimental outcomes reveal that the proposed approach is better in energy cost saving than the existing algorithms in DCs.

Blockchain presented a resource management system termed BCEdge in D2D-assisted mobile edge computing [20]. This system discharges the load of edge clouds and is a reliable scheme. Using interaction charts and flow charts, the advantages and technical details of BCEdge are explained in the study. The experiment results confirm the benefits of the proposed scheme.

A resource allocation algorithm has been presented [21], which employs hybrid differential parallel scheduling. In this approach, the authors first designed the data and the grid structure. The resource attributes are classified using the clustering analysis process; then the sliding window is split into multiple subwindows. Employing singular value decomposition, the resources are allocated. As a least squares problem, the resource allocation problem is transformed. An optimal solution is obtained using this proposed approach to find the resource scheduling vector set. The experimental outcomes showed the effectiveness of this approach.

Praveenchandar et al. [22] presented a hybrid optimization algorithm using simulated annealing (SA) and artificial bee colony (ABC). Efficient scheduling is done based on the priority of the request, the size of the task, and the optimal distance. The proposed approach improves the ability based on the optimal resource time where the SA algorithm executes the dynamic and random searching process. Using the CloudSim tool, the implementation process is done. The experimental results were compared with the existing system.

The same authors [23] conducted a study on effectual dynamic resource allocation. This work concentrates on the power efficiency and enhanced scheduling of tasks. Using the dynamic resource table and prediction mechanism, the efficiency of the proposed method is achieved. This methodology achieves improved results based on power minimization as it reduces power depletion. The resource table is updated with the accurate values provided by the proposed approach. This approach achieves efficient resource management using a better task scheduling technique with minimum power consumption.

An effective algorithm for task scheduling has been presented [24]. The tasks of the user in this approach are stored in the queue manager. This approach calculates the priority, and a proper resource allocation is created for a task. Energy consumption is still required for the 5G wireless platform to develop a smart-grid system [25]. Even 6G is also used for low latency and better efficiency in comparison to 5G.

6.3 BACKGROUND

This section presents an overview of several aspects of cloud and blockchain technology. The aspects are privacy and security concerns, resource allocation model, performance evaluation, QoS, etc.

6.3.1 Overview on Cloud Computing

According to the National Institute of Standards and Technologies (NIST), cloud computing is a paradigm for empowering pervasive, appropriate, on-demand network support to a shared pool of configurable computing to which delivery of these resources can be done rapidly with the least contact with the provider and with very less administrative effort. Based on the offered services and location, cloud computing can be categorized into two means: A cloud can be private, public, community, or hybrid based on the location [26]. A private cloud offers a good level of control to the user at a higher cost and with improved security features; moreover, it is available to an organization or a specific user. Anyone can access public cloud services, and a service-providing company holds its infrastructure. They are prone to several attacks and are very economical. Many organizations use the community cloud, which has a communal structure that possesses shared data and management. Cloud classification is mainly done based on services such as infrastructure as a service (IaaS), platform as a service (PaaS), or software as a service (SaaS) [27]. The public and private clouds are combined to form the hybrid cloud used for various rationales based on the organization's requirements. The cloud in IaaS offers elementary IT resources such as computers, networking features, more flexibility, and computing resource control. The organization typically needs to deal with the basic infrastructure, computer hardware, and operating system (OS) that are taken over by PaaS, enabling it to concentrate on the deployment of applications. Users are allowed to focus on specific software by SaaS free of judging how the services and infrastructures are managed. Cloud computing, along with these services, offers many services such as storage as a service (SaaS), expert as a service (EaaS), communication as a service (CaaS), database as a service (DaaS), monitoring as a service (MaaS), security as a service (SECaaS), testing-as-a-service (TaaS), and network as a service (NaaS), which are used for several application goals by the user [28]. Many available cloud applications are available for various aspects of robotics, educational purposes, data analysis, and health monitoring. Lacking the knowledge of the required technology, users can access them for their computing environment. As the cloud provides many services, they require QoS monitoring to handle the services offered to meet users' demands and sustain the SLA.

The cloud comes over several issues during this process: privacy and security issues, load balancing, modeling and performance analysis, resource management, response time, QoS, and throughput. For a service provider, resource allocation is a complication due to heterogeneous applications (intensive memory and CPU), demand requests, locality limits, and resource heterogeneity in the cloud environment. In the cloud and 6G-oriented environments, effective resource allocation has a vital role in meeting the demands of service providers and users. Thus, to deal with cloud users' intermittent demands, a dynamic resource allocation algorithm is mandatory [29–31].

6.3.2 Overview of Blockchain Technology

Blockchain is termed a collective repository of archives or a source of contracts (digital events) that are accomplished and distributed among the participating entities. The data in the blockchain is kept in a similar form as the distributed database, and its designs make manipulation very difficult as the participants of the network save and verify the blockchain. Every block is constructed with a body and a header. Using the index technique, the data is searched in the database. The hash values of the current and previous blocks are present in the header, including the nonce [32]. While the hash value of the next block is included as a practice, the block does not contain it. As the stored hash values in the block are affected by the values of the previous block, it is a complex subject to alter and falsify the registered data. Simultaneously, if 51% of peers are hacked, there is a possibility of data alteration if there is a critical attack scenario. In the blockchain, a hash function and key-based verification are used to provide security. The digital signature produced in a transaction between individuals is verified. There is no alteration in the data transferred.

The public key-based encryption is used to verify the transaction integrity by the transaction data's hash value. Using these features of blockchain, several research works are ongoing to strengthen security. The vital part of blockchain is the personal key-based security used in encryption [33]. Bitcoin can be hacked by the attacker, which might cause the data to be dripped, causing the attacker to attain the personal key. To deal with this issue, research works are carried out using securities relying on software and hardware. The malware easily affects Bitcoin as it is susceptible to infections owing to its applications in commonly used devices such as smartphones or PCs. Proper treatment should be applied to the malware affecting facilities such as USB, e-mail, or poor security applications since a peer's device can also be infected with it. As many of them use bitcoins, the need for security is also rising, mainly in gaming items. In the game environment, there are works on treating and detecting malware [34].

The main advantage of bitcoin is that it is impossible to alter and falsify the ledger as many peers share the transaction ledger. In most ledgers, it takes the recorded data; besides, hacking is almost impossible unless the attacker falsifies authorization and alters 51% of all the ledgers of the peers, tampering with some of the ledger's data [35]. In the public register, every transaction is validated by the consent of a majority of the system's participants. The information, once entered, cannot be deleted. The record of all the transactions ever made is preserved in the block. The blocks are of three types:

Public: Bitcoin, for example, runs through a native symbol and is considered a large distributed network. Anyone can take part as the blocks are open for all and comprise an available open-source code.

Permission: For example, Ripple organizes roles that entities can play on the network. Still, native tokens are used by distributed and large systems. The code of them may or may not be open source.

Private: Private blocks do not use a token and are smaller. They have control of their belongings. The consortiums used in these blocks are helpful in confidential trading information.

Cryptography is used in all types of blocks and permits every entity of a network to oversee the registry safely, independently of the need for a principal hub [36]. One of the powerful and vital aspects of the blocks is eliminating the principal hub from the database structure. An administrator is not needed as the users of the blockchain will be the administrators. For "smart contracts" or scripts, these networks can be used, which automatically function when certain criteria are met. Preestablished conditions are met by Ethereum users that show someone owns a certain token and possesses the power to transfer the money they own. Multiple blockchain users spawn contracts and, to spark a transaction, they need more than one set of entries; i.e., signatures are required among the sellers, buyers, and their financial institutions for real estate transactions [37].

6.3.3 Blockchain vs. Cloud

The predecessors of blockchain are the cloud paradigm, and the opportunities of blockchain should also be considered by developers and inventors in this domain. In cloud security environments, the private networks of blockchain can also be run, which leads to taking part in the main role of blockchain implementation. The blockchain and cloud at the same time have several problems in common, but the utilization of cloud blockchain pilot projects could provide many benefits. The security protection systems of the cloud and blockchain permit the encryption of data. The cloud deployment model's capacity to clear the private address, community, and public states impeccably fit with the properties of the block, determined by the design members of the chain. These are unaffected by cybercrime. Due to the peer-topper model, the cloud is secured from aggressively contributing and professionally evolving toward the formation of cyber-free areas [38]. This includes proactive identification and 24-hour monitoring for mistrustful deeds with real-time response to such threats. Blockchain and cloud considerably reduce expenses. Possible design faults are prevented by the blockchain, along with the cloud, given its processes. The storage infrastructure before the arrival of cloud service providers was very decentralized.

On their premises, the cloud service providers own the servers, leading to high expense but had great control. The organization needs internal expertise to maintain and configure costly servers and units. Service providers such as Amazon S3, at an estimated 25 USD per terabyte a month, provides a tremendously valuable, reliable, and time-redundant service by replicating data across multiple data centers. This facility has led us to ignore the drawbacks. When using a cloud service, a lot of confidence is placed in the third parties, who are trusted to safeguard our private and most sensitive data [39]. Usually, this data is unencrypted, and, due to the lack of a good point of reference, the expense is much lower. A distributed and decentralized storage market is created with the aid of blockchain. The most successful markets meet growing demands with insufficient resources. In the block storage market, excess hosts sell their storage capacity, the tenants acquire this surplus capacity, and the files are uploaded. In dozens of nodes, the files are fragmented, encrypted, and distributed intelligently. Certain things are permitted using a formerly difficult approach. The Amazon S3 performs redundancy by allocating files to regional data centers, which tend the data center, a significant point of failure. The user data is governed by no third party with access to user files. The nodes store only encrypted user data fragments.

6.3.4 Categories of Blockchain

This section briefly presents the blockchain types. In the blockchain, there are two types of ledger:

(A) Private or centralized ledger or (B) public or decentralized ledger. Blockchain is of two types based on permission: (C) permissioned and (D) permission-less.

A. *Public or decentralized*: This particular ledger is based on proof of work consensus algorithms that are not permission (pp) and source. Any person can write, read, and send information. The buyer creates the block or the transaction. Later, with the help of cryptographic hashing, the transaction is distributed or validated—for example, Ethereum, Monero, Bitcoin, Dogecoin, Litecoin, Dash, etc.

B. *Private or centralized*: In this particular ledger, the permission to write is consolidated to one establishment while the permission to read is restricted or available publicly. The benefits of blockchain technology are grabbed by end users when they verify the transactions. In the present situation, these types of ledgers are comparatively outdated.

C. *Permissioned*: In this network, every node is supported in the consensus method. This blockchain might be private-permissioned or public-permissioned. In private-permissioned, the transactions are validated by the consortium member with the PBFT (practical Byzantine fault tolerance) or multisignature.

D. *Permission-less*: This particular blockchain has an apparent and public proprietorship, which functions on proof of work. Anyone can validate transactions and access the protocol [40].

6.3.5 Blockchain Platforms

In a permission-less blockchain, Ethereum, as an example, works on PoW (Ethash) consensus appliance and uses ether currency. When it comes to permissioned blockchain technology, the hyperledger is a type that works on a PBFT (excluding Corda) appliance and doesn't use any currency. Still, employing chain code in the hyperledger using inherent currency is feasible. For any application, Ethereum prevails as a generic platform. Its permission-less mode and flexible nature come at the price of privacy, scalability, and performance. The hyperledger addresses these issues by the application's permission model. Corda can be adopted in abundant applications and is sectional in structure.

Primarily, Corda emphasizes the transaction of financial services [41]. To handle double-spending and various types of attacks, machine learning is also used to facilitate decision-making conditions. Blockchain and data science can handle security threats for smart applications.

6.3.6 Applications of Blockchain

A. *Digital identity*: More control of personal data is given by the blockchain digital identity, which recognizes the proprietors and lessens the company disquiet—it verifies to work cohesively. It identifies the issues on the same platform. It builds a decentralized structure wherein blockchain activates the point-to-point exchange of information. The code of conduct is followed by the issuer, identity holders, and auditors.

B. *Smart contracts*: The skeptical groups in the decentralized cryptographic epoch claim to act safely for third parties. As a solution, smart contracts evolved, which issue confirmation for fair compensation, which the parties pay. The functioning of smart contracts is depicted in Figure 6.1.

C. *Registry*: For specific objects, the registry is an influential database. The public authorities organize and manage many public registers, confirming the validity of the registered objects. Any variation in the logbook is noted with a digital fingerprint, which is freely managed. Every change must be noted in the register and also should be open to independent assessment. To reduce errors and duplicate records, a register can be referred to other records.

6.3.7 Resource Management Using Blockchain

This section discusses in brief administering the requests and virtual machines among cloud data centers using blockchain.

Preliminary

All members of a blockchain network are a linked data structure. How to handle the consensus issues of the bitcoin network was presented by S. Nakamoto. The structure of blockchain is illustrated in Figure 6.2. Figure 6.3 shows various applications of blockchain in the development of smart city.

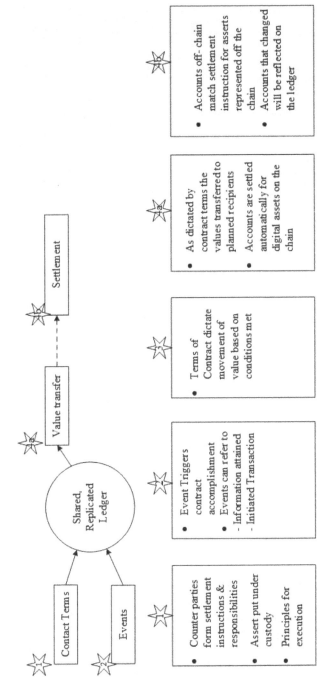

Figure 6.1 Functioning of smart contracts.

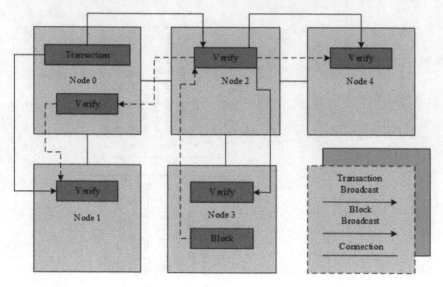

Figure 6.2 Structure of blockchain.

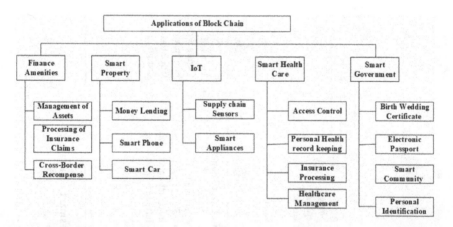

Figure 6.3 Blockchain applications.

As a single list, the blockchain is organized, in which all blocks, except the first block (genesis block), contain the hash of the previous block. Figure 6.4 presents a detailed description of the generation mechanism of the blockchain. The description of Figure 6.4 shows that:

1. A transaction is signed by the user using the private key while interrelating with Node0. Thus, using the user's public key, the transaction can

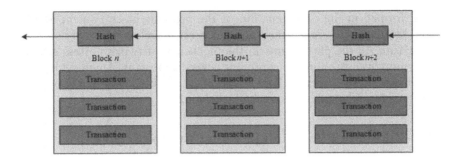

Figure 6.4 Blockchain network.

be traced. Then, to the one-hop neighbor of Node0 (i.e., Node1 and Node2), the transaction is made.

2. The broadcast transaction is verified by the neighboring nodes (i.e., Node1 and Node2), obeying the transaction protocol and preventing the failing of the transaction broadcast to neighbors (Node3 and Node4). In the design of the blockchain, the determination of the transaction protocol is also done. The transaction protocol in the blockchain network prevents chaos. This transaction spreads throughout the entire blockchain network by repeating these procedures.

The network generates the transactions by all participants at a fixed time interval using a mining node that is packaged into one block. Then:

1. The block to the blockchain network is broadcasted by the miner (i.e., Node3). The block is also broadcast peer to peer like the transaction. As a result, the block is received by Node2.
2. The Node2, which is the receiver, validates:

   a. Whether all the transactions of the block obey the transaction protocol.
   b. In the blockchain, a correct hash is present in the block of the previous block.

The receiver adds it to the blockchain once the verification is passed by the block and extracts the block containing the transactions to update the receiver's transactions, which is known as the "view of the world." Or else the block is rejected. In the blockchain network, the consensus mechanism leverages the miner's choice.

To propose the next block of the blockchain in bitcoin, the random value of the nodes first found should be entitled. The attained value must produce a number less than a given threshold once the double SHA-256 hash operations are performed in the concatenation of the previous block content and the value. Mining is also called the finding process, which is a sort of consensus mechanism. All the participants in the blockchain network possess a blockchain; moreover, the consensus process is very important. Anyone who commonly uses proof of stake (PoS) or proof of work (PoW) as its consensus process can use the public network blockchain.

6.3.8 Resource Allocation in Cloud Computing

The allotment of the resources of the cloud provider to a customer is said to be cloud provisioning.

When a cloud provider accepts a request from the customer, a suitable number of VMs (virtual machines) are created, and the resources are allocated to aid them. Likewise, dynamic provisioning, user self-provisioning, and advance provisioning take the lead. The phrase "provisioning" denotes "to provide." The customer with new provisioning assures the provider of service and appropriate resources are prepared at once by the provider to initiate the service. On a monthly basis, the customer is charged or billed with a tariff, based on pay per use. When dynamic provisioning is employed to create a hybrid cloud, it is described as cloud bursting.

Figure 6.5 depicts the flowchart for resource allocation. The service request is submitted by the cloud users and from anywhere with a web interface or graphical user interface. Many data centers are accessible in a cloud environment to process the request, but due to low latency, the request is steered to the nearest data center. If the request is not directed to the nearest data centers, it might lead to complexions that affect some QoS parameters such as response time, deadline, etc.

When the QoS parameters are affected, there is an increase in the SLA violation. The gatekeeper or the job request handler receives a user service request at the data center. A Turing test is conducted in the data center to confirm that the request is coming from an attacker or a user. If it is from an attacker, then the Port address, IP of the source, is blocked. Or else the request succeeds to the controller node that has the sensitive information. The controller node checks the availability of cloud and 6G resources. The available resources are selected on behalf of the QoS parameter like a deadline, makespan, response time, etc. They are stopped for resource availability.

6.4 PROPOSED RESEARCH APPROACH

This section explains the benefits offered by the cloud from the technical, operational, and economic viewpoints. An improvement is needed for allocating the resource in the data centers and smart grid for the cloud and 6G platform, respectively.

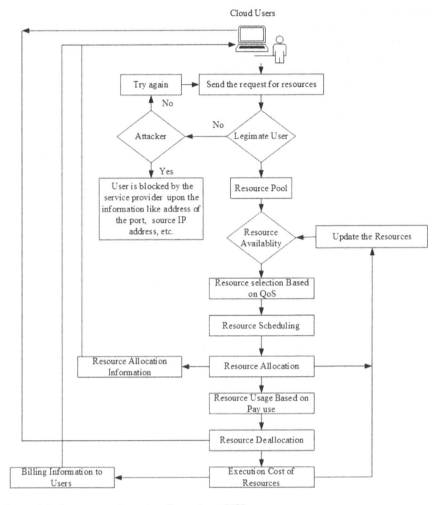

Figure 6.5 Resource allocation flow chart [42].

In this study, the proposed approach highlights the vital complications encountered by the providers of cloud infrastructure. Resource management is mainly done using a blockchain-based framework. The main intention of adopting this framework is to save the expense on energy in the existing models. For effective resource management, some of the factors considered are SLA dynamic resource demand pattern and resource utilization. The main objective is to effectively allocate the resources with minimum makespan to achieve an energy-efficient schedule. Many decision-making problems and real-world designs involve real-time optimization of multiple objectives. An optimization model is designed for resource allocation that will fully incorporate the preceding two energy efficiency factors and makespan optimization in this work. The SMO is used to optimize resource allocation based on makespan, and a brown-out based energy model is adopted to reduce the reduce energy and carbon footprint.

6.4.1 Spider Monkey Optimization Process

The observation and intentions of spider monkeys are identified by the positions and postures at long distances. By particular sounds such as chattering or whooping, they interact with one another. Every monkey has its unique noticeable sound by which other group members identify a monkey. Figure 6.6 depicts the spider monkey's foraging behavior.

This algorithm is recognized as a metaheuristic method. The concept of this algorithm is the social behavior of spider monkeys, which implements the swarm intelligence method for foraging based on fission and fusion. Swarm is the living place of the spider monkeys, comprising 40–50 members. In a region, the leader decides to divide the strategy for food searching. Female leads create mutable small clusters and are the leads of the swarm. The cluster size depends on the availability of food sources as well as the region. The essential criterion of the SMO algorithm is swarm intelligence (SI), comprising labor division and self-organization. The foraging work is divided by the spider monkeys' labor division by creating smaller groups. To meet food availability, the principle of self-organization is followed. Thus, an SMO-based algorithm is divided into a normal and swarm- and intelligence-based algorithm.

6.4.2 Implementation of SMO Algorithm

Implementation of this algorithm is done using six-stages: local leader stage, local leader learning stage, local leader decision stage and global leader stage, global leader, learning stage, and global leader decision stage.

6.4.2.1 Population Initialization

The population of and SMO is uniformly distributed. The spider monkeys are denoted as SMP. In which $p = 1, 2 \ldots P$ and in the population of the pth monkey, which is signified as SMP. As M-dimensional vectors, the monkeys are identified, where M determines the sum of problem domain variables. Each spider monkey is related to an optimal solution for the problem. SMO uses the following equation to initialize every SMP:

$$SM_{pq} = S_{Mminq} + UR(0,1) \times (SM_{maxq} - S_{mminq}) \tag{6.1}$$

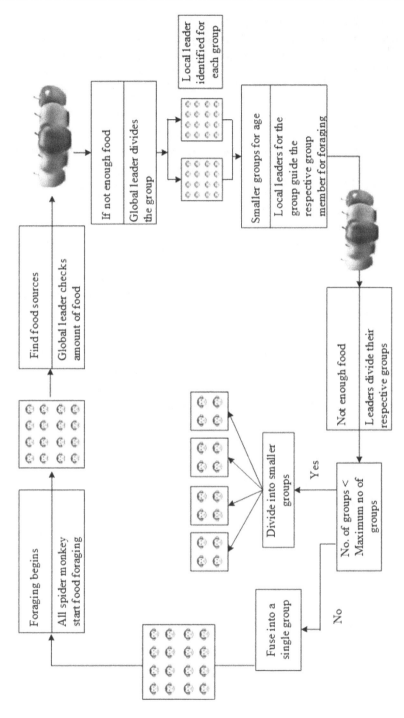

Figure 6.6 Foraging behavior of spider monkeys [43].

where:
SM_{pq} denotes the qth dimension of the pth spider monkey;
SM_{minq} denotes the lower bound of SM in the qth direction;
SM_{maxq} denotes the qth direction upper bounds of SM_p (where q =1, 2. . . M);
A random number is denoted by UR(0,1) distributed over the range of [0, 1].

6.4.2.2 Local Leader Stage (LLS)

The next step is LLP. It uses the past events of both local leaders and local group members to modify SM's current location. The position of the SM is only upgraded to a new spot whenever the current location's fitness value is preferable to the former one. The local group expression for the pth SM is shown in Eqn. (6.2):

$$SM_{newpq} = SM_{pq} + UR(0, 1) \times (LL_{lq} - SM_{pq}) + UR(-1,1) \times (SM_{rq} - SM_{pq}) \quad (6.2)$$

The qth length of the lth local group leader location is denoted as LL_{lq}, the qth length of the arbitrarily chosen lth SM of the lth local group is denoted as SM_{rq} meets the condition that $r \neq p$.

6.4.2.3 Global Leader Stage (GLS)

The third step of the implementation is GLS. In this step, the experience of global leaders and the new location of SM are upgraded using the local group members. The SM location of the equation is given in Eq. (6.3):

$$SM_{newpq} = SM_{pq} + UR(0, 1) \times (GL_{lq} - SM_{pq}) + UR(-1,1) \times (SM_{rq} - SM_{pq}) \quad (6.3)$$

The location of a global leader in qth dimension is denoted in Eq. (6.3), where GL_q and the q values are assigned as 1, 2, 3, the arbitrarily designated index is termed as M. Fitness value of SM is utilized to determine prb_p (probability), and the location is also updated in such a way. The members at the proper position have access to more chances to develop. The estimation of probability equation is:

$$prb_p = \frac{fnp}{\sum_{p=1}^{N} fn_p} \quad (6.4)$$

The fitness value of the pth SM is denoted as *fnp*. Then the new location's fitness value is assessed and compared with the previous location. Finally, the preeminent fitness value is chosen for further processing.

6.4.2.4 Global Leader Learning Stage (GLS)

The global leader position is updated by adopting the greedy selection method. The position of the SM with the maximum fitness value in the community is updated to the leading global location. The global leader provides the best position. If there are no additional updates, the global limit count tends to add an increment of 1.

6.4.2.5 Local Leader Learning Stage (LLS)

The greedy search approach all through the local group is used to upgrade the local leader position. *SM* and the best fitness update the location of a local leader in a particular local group. The local leader has an optimal location. If no additional updates are found, then the count of local limit increased by 1.

6.4.2.6 Local Leader Decision Stage (LLS)

Within the limit of a local leader, if the local leader is not updated, then the candidates of the local group, as per step 1or by using the previous data from the local leader and the global leader, modify their locations, through the following Eqn. (6.5).

$$SM_{newpq} = SM_{pq} + \mathrm{UR}(0,1) \times (GL_{lq} - SM_{pq}) + \mathrm{UR}(0,1) \times (SM_{rq} - LL_{pq}) \quad (6.5)$$

6.4.2.7 Brownout-Based Energy Model

In data processing and migration, the real concern is energy, which is measured by the power utilized by each VM for the smart grid system. The cloud domain energy mainly relies on the power expended from several resources of the VM. This brownout mechanism is adopted in this proposed model to reduce energy and carbon footprint. Resource usage is controlled by this approach by actively controlling the applications to microservices.

6.5 EXPERIMENTAL SETUP AND RESULTS

This section validates the efficiency of the approach proposed. The analyzed factors are execution time, acceptance ratio, resource exploitation, and power management.

The cloud sim software is used for simulation. The cloud data center is comprised of several PM's. The resource agents in the data centers are initiated. A series of hosts with their equivalent VMs are initiated in each data center. The cloudlet scheduler schedules the incoming tasks. The hardware requirements are displayed in Tables 6.1 and 6.2.

Table 6.1 Hardware Requirements

Resources	Specifications
Processor	Intel Pentium CPU G2030@3.00GHZ
Hard disk	1TB
RAM	4GB
OS	Windows (X86 ultimate) 64 bit

Table 6.2 Simulation Parameters

Resources	Specifications	Values
VM	Host	4
Cloudlets	Length of task	1600–3400
	Number of tasks	30–3000
Physical machine	Bandwidth	250,000
	Memory	540
	MIPS/PE	500
	Storage	500 GB

The efficiency of the system is enhanced by the proposed approach and is compared with the general approaches.

The graphical representation reveals the performance of the proposed approach. For calculating the resource utilization, the ratio of the consumed resources in a data center is analyzed. The tasks arrive as a batch in this approach. For job arrival, the batch processing concept is borrowed. The proposed approach is used for the allocation of a task. The task scheduling concept is applied to prioritize the jobs. The tasks that are prioritized are given to the agents for allocating the available resources. Some of the existing approaches GA [44], PSO [43], and ACO [45] are used for comparison during this process, and the parameters calculated are bandwidth, acceptance ratio, execution time. The further suggested related works are also in [46–51].

6.5.1 Performance Evaluation

The performance evaluation is done with certain metrics, i.e., resource utilization, makespan, task completion ratio, execution time, and power consumption.

6.5.1.1 Task Response Time (TRes)

Time taken from the arrival of the task to the execution of the task is called the response time of a task. Let the task complete-time be T_{Cmp}, and

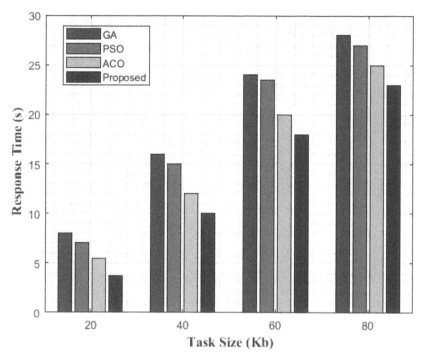

Figure 6.7 Comparison of response time.

the task arrival time be T_{Start}. The response time T_{Res} can be denoted as follows:

$$T_{Res} = T_{Cmp} - T_{Start} \qquad (6.6)$$

The response time, as seen in Figure 6.7, is much less when compared to the other methods.

6.5.1.2 Make Span (M_{span})

The execution time taken for a set of tasks as of the start to the end of the task is called makespan. The time taken to accomplish all tasks is T_C. The M_{span} is denoted subsequently.

$$M_{span} = Max(T_C) \qquad (6.7)$$

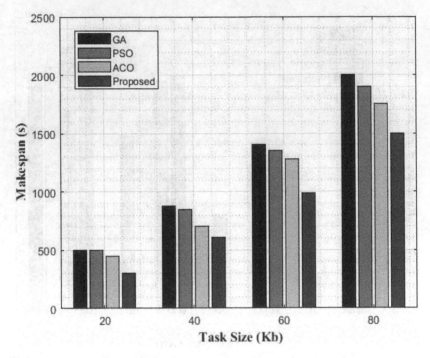

Figure 6.8 Comparison of makespan.

Minimized makespan is attained when compared with the existing algorithms. Graphical representation in Figure 6.8 shows a reduction in makespan and a rise in efficiency of about 30% over existing methods.

6.5.1.3 Resource Utilization (R_U)

The volume of resources available is exploited by the tasks to be fulfilled; this is called resource utilization. The available resources are represented as R_{avl}, and the unused resourced is represented as R_{un}. Resource utilization (R_U) is denoted as:

$$R_U = R_{avl} - R_{un} \tag{6.8}$$

Both memory and CPU are used in the proposed work. The resource utilization percentage of the proposed is higher when compared with the existing algorithms.

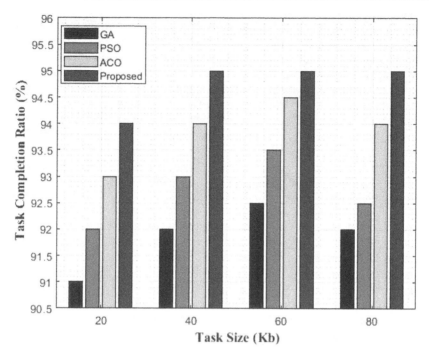

Figure 6.9 Comparison of resource utilization.

The maximum utilization of resources is displayed in Figure 6.9. The proposed approach puts idle PMs in the OFF state.

6.5.1.4 Task Completion Ratio (TCR)

TCR is a ratio between the effectively accomplished tasks to the submitted tasks at a definite time. The number of completed tasks is denoted as N_{scc}, and the submitted tasks are N_{sbt}. The completion ratio of the task is denoted:

$$T_{CR} = N_{scc} \Big/ N_{sbt} \qquad (6.9)$$

In the simulation, the completed tasks for a definite period are analyzed. The improved performance ratio is attained from the simulation results. The proposed approach is compared with the existing algorithms for validating the performance.

The graphical representation in Figure 6.10 reveals the dominance of the presented approach when compared with the other.

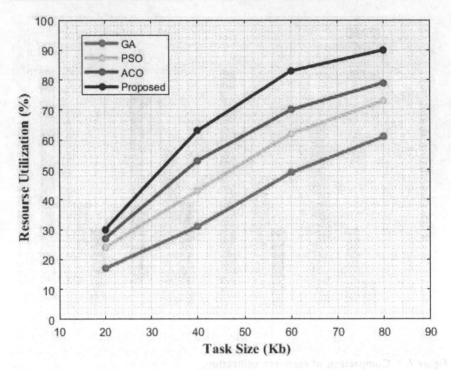

Figure 6.10 Comparison task completion ratio.

6.5.1.5 *Power Consumption*

The power used during the resource allocation process by the PMs is defined as power consumption for both the cloud and the 6G platforms. It is clear as all PMs utilize the total power. The brownout approach is adopted in this proposed work to minimize energy utilization. In real-time data centers, many technologies deal with power consumption, such as resource hibernation, dynamic voltage scaling, and dynamic voltage frequency scaling, which don't make sense in the virtualized environment. The implemented approach for power minimization in this study is superior when compared to existing methods.

Figure 6.11 shows that the proposed approach consumes minimum power, less than existing methods.

6.5.2 Major Findings

The major findings are that there is a need for a technique that fulfills the requirements in terms of the parameters of the blockchain-based resource

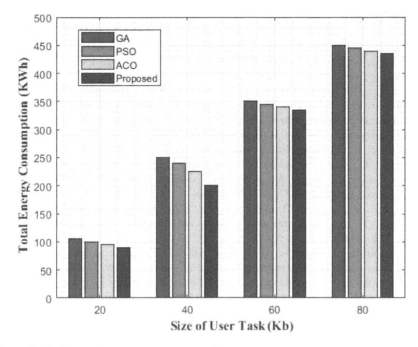

Figure 6.11 Comparison power consumption.

management with energy efficiency for cloud and 6G topology. The considered parameters in this chapter are: (1) optimum resource utilization, (2) minimization of the makespan's inefficient manner, (3) task completion ratio, (4) minimization of energy consumption, and (5) reducing optimization of response time.

6.6 CONCLUSIONS AND FUTURE WORK

In this proposed work, we conclude the simulation outcomes of makespan, resource utilization, task completion ratio, minimizing energy utilization, and comparison of response time efficiently for a cloud data center and intelligent smart grid environment for 6G communication. This is a vital concern in this domain, especially when dealing with the cloud of a large-scale structure. The interarrival times and the task sizes in the proposed blockchain-based resource management and SMO-based energy-efficient resource allocation mechanism are considered. The proposed mechanism in the future will be parallelized to reduce the time to find solutions.

REFERENCES

1. Yang, J., Z. Xiang, L. Mou, and S. Liu.: Multimedia resource allocation strategy of wireless sensor networks using distributed heuristic algorithm in cloud computing environment. *Multimedia Tools and Applications*, pp. 1–15, 2019.
2. Strinati, E.C., S. Barbarossa, J.L. Gonzalez-Jimenez, D. Ktenas, N. Cassiau, L. Maret, and C. Dehos: 6G: The next frontier: From holographic messaging to artificial intelligence using subterahertz and visible light communication. *IEEE Veh. Technol. Mag.*, 14(3) pp. 42–50, 2019.
3. Priyaa, A. R., E. R. Tonia, and N. Manikandan.: Resource Scheduling Using Modified FCM and PSO Algorithm in Cloud Environment. In *International Conference on Computer Networks and Inventive Communication Technologies*. Springer, Cham. pp. 697–704, 2019.
4. Shooli, R. G., and M. MasoudJavidi. Using gravitational search algorithm enhanced by fuzzy for resource allocation in cloud computing environments. *SN Applied Sciences*, 2(2), p. 195, 2020.
5. Varshney, S., R. Sandhu, and P. K. Gupta.: QoS based resource provisioning in cloud computing environment: A technical survey. In *International Conference on Advances in Computing and Data Sciences*. Springer, Singapore, pp. 711–723, 2019.
6. Suresh, A., and R. Varatharajan.: Competent resource provisioning and distribution techniques for cloud computing environment. *Cluster Computing*, pp. 1–8, 2019.
7. Rekha, P. M., and M. Dakshayini.: Efficient task allocation approach using genetic algorithm for cloud environment. *Cluster Computing*, 22(4), pp. 1241–1251, 2019.
8. Devarasetty, P., and S. Reddy.: Genetic algorithm for quality of service based resource allocation in cloud computing. *Evolutionary Intelligence*, pp. 1–7, 2019.
9. Durgadevi, P., and S. Srinivasan.: Resource allocation in cloud computing using SFLA and Cuckoo search hybridization. *International Journal of Parallel Programming*, 48(3), pp. 549–565, 2018.
10. Xavier, V. M. A., and S. Annadurai.: Chaotic social spider algorithm for load balance aware task scheduling in cloud computing. *Cluster Computing*, 22(1), pp. 287–297, 2019.
11. Prassanna, J., and N. Venkataraman.: Threshold based multi-objective memetic optimized round Robin scheduling for resource efficient load balancing in cloud. *Mobile Networks and Applications*, 24(4), pp. 1214–1225, 2019.
12. Valarmathi, R., and T. Sheela.: Ranging and tuning based particle swarm optimization with bat algorithm for task scheduling in cloud computing. *Cluster Computing*, 22(5), pp. 11975–11988, 2019.
13. Panwar, N., S. Negi, M. M. S. Rauthan, and K. S. Vaisla.: Topsis—pso inspired non-preemptive tasks scheduling algorithm in cloud environment. *Cluster Computing*, 22(4), pp. 1379–1396, 2019.
14. Sharma, P. K., N. Kumar, and J. H. Park.: Blockchain technology toward green IoT: Opportunities and challenges. *IEEE Network*, 34(4), pp. 263–269, July/August 2020.
15. Tuli, S., R. Mahmud, S. Tuli, and R. Buyya. Fogbus: A blockchain-based lightweight framework for edge and fog computing. *Journal of Systems and Software*, 154, pp. 22–36, 2019.

16. Gill, S. S., S. Tuli, M. Xu, I. Singh, K. V. Singh, D. Lindsay, and S. Tuli. Transformative effects of IoT, Blockchain and: Artificial intelligence on cloud computing: Evolution, vision, trends and open challenges. *Internet of Things*, 8, pp. 100–118, 2019.

17. Wang, H., L. Wang, Z. Zhou, X. Tao, G. Pau, and F. Arena.: Blockchain-based resource allocation model in fog computing. *Applied Sciences*, 9(24), pp. 5538, 2019.

18. Xu, C., K. Wang, and M. Guo.: Intelligent resource management in blockchain-based cloud datacenters. *IEEE Cloud Computing*, 4(6), pp. 50–59, November/December 2017.

19. Yu, C., L. Zhang, W. Zhao, and S. Zhang. A blockchain-based service composition architecture in cloud manufacturing. *International Journal of Computer Integrated Manufacturing*, 33(7), pp. 701–715, 2020.

20. Zhou, A, Q. Sun, and J. Li.: BCEdge: Blockchain-based resource management in D2D-assisted mobile edge computing. *Software: Practice and Experience*, 51, 2019.

21. Wei, J., and Xin-fa Zeng.: Optimal computing resource allocation algorithm in cloud computing based on hybrid differential parallel scheduling. *Cluster Computing*, 22(3), pp. 7577–7583, 2019.

22. Muthulakshmi, B., and K. Somasundaram.: A hybrid ABC-SA based optimized scheduling and resource allocation for cloud environment. *Cluster Computing*, 22(5), pp. 10769–10777, 2019.

23. Praveenchandar, J., and A. Tamilarasi.: Dynamic resource allocation with optimized task scheduling and improved power management in cloud computing. *Journal of Ambient Intelligence and Humanized Computing*, pp. 1–13, 2020.

24. Kumar, A. M. S., and M. Venkatesan.: Task scheduling in a cloud computing environment using HGPSO algorithm. *Cluster Computing*, 22(1), pp. 2179–2185, 2019.

25. Joda, R. et al.: Carrier aggregation with optimized UE power consumption in 5G. *IEEE Networking Letters*. doi: 10.1109/LNET.2021.3076409.

26. Narman, H. S., M. Hossain, M. Atiquzzaman, and H. Shen.: Scheduling internet of things applications in cloud computing. *Annals of Telecommunications*, 72(1–2), pp. 79–93, 2017.

27. Masdari, M., S. S. Nabavi, and V. Ahmadi.: An overview of virtual machine placement schemes in cloud computing. *Journal of Network and Computer Applications*, 66, pp. 106–127, 2016.

28. Singh, S., and I. Chana.: A survey on resource scheduling in cloud computing: Issues and challenges. *Journal of Grid Computing*, 14(2), pp. 217–264, 2016.

29. Khaleel, M., and M. M. Zhu.: Energy-efficient task scheduling and consolidation algorithm for workflow jobs in cloud. *International Journal of Computational Science and Engineering*, 13(3), pp. 268–284, 2016.

30. Kumar, M., and S. C. Sharma.: PSO-COGENT: Cost and energy efficient scheduling in cloud environment with deadline constraint. *Sustainable Computing: Informatics and Systems*, 19, pp. 147–164, 2018.

31. Sharma, H., G. Hazrati, and J. C. Bansal.: Spider monkey optimization algorithm. In *Evolutionary and Swarm Intelligence Algorithms*. Springer, Cham, pp. 43–59, 2019.

32. Zheng, Z., S. Xie, H. Dai, X. Chen, and H. Wang.: An overview of blockchain technology: Architecture, consensus, and future trends. In *2017 IEEE International Congress on Big Data (BigData Congress)*. IEEE, Honolulu, Hawaii, pp. 557–564, 2017.

33. Wu, J., and N. K. Tran.: Application of blockchain technology in sustainable energy systems: An overview. *Sustainability*, 10(9), p. 3067, 2018.

34. Ahram, T., A. Sargolzaei, S. Sargolzaei, J. Daniels, and B. Amaba.: Blockchain technology innovations. In *2017 IEEE Technology & Engineering Management Conference (TEMSCON)*. IEEE, Santa Clara, CA, pp. 137–141, 2017.

35. Sheetal, M., and K. A. Venkatesh.: Necessary requirements for blockchain technology and its applications. *International Journal of Computer Engineering and Information Technology*, 9, pp. 130–134, 2018.

36. Zhang, L.: Blockchain: The new technology and its applications for libraries. *Journal of Electronic Resources Librarianship*, 31(4), pp. 278–280, 2019.

37. Saberi, S., M. Kouhizadeh, J. Sarkis, and L. Shen.: Blockchain technology and its relationships to sustainable supply chain management. *International Journal of Production Research*, 57(7), pp. 2117–2135, 2019.

38. Xu, C., K. Wang, and M. Guo.: Intelligent resource management in blockchain-based cloud datacenters. *IEEE Cloud Computing*, 4(6), pp. 50–59, 2017.

39. Xie, S., Z. Zheng, W. Chen, J. Wu, H.-N. Dai, and M. Imran.: Blockchain for cloud exchange: A survey. *Computers & Electrical Engineering*, 81, p. 106526, 2020.

40. Silvestre, D., M. Luisa, P. Gallo, J. M. Guerrero, R. Musca, E. R. Sanseverino, G. Sciumè, J. C. Vásquez, and G. Zizzo.: Blockchain for power systems: Current trends and future applications. *Renewable and Sustainable Energy Reviews*, 119, p. 109585, 2020.

41. Maesa, D. D. F., and P. Mori.: Blockchain 3.0 applications survey. *Journal of Parallel and Distributed Computing*, 138, pp. 99–114, 2020.

42. Tanwar, S., Q. Bhatia, P. Patel, A. Kumari, P. K. Singh, and W. Hong.: Machine learning adoption in blockchain-based smart applications: The challenges and a way forward. *IEEE Access*, 8, pp. 474–488, 2020.

43. Kumari, A., A. Shukla, R. Gupta, S. Tanwar, S. Tyagi, and N. Kumar.: ET-DeaL: A P2P smart contract-based secure energy trading scheme for smart grid systems. In *IEEE INFOCOM 2020*. IEEE Conference on Computer Communications Workshops (INFOCOM WKSHPS), Toronto, ON, pp. 1051–1056, 2020.

44. Tanwar, S., S. Kaneriya, N. Kumar, and S. Zeadally.: ElectroBlocks: A blockchain-based energy trading scheme for smart grid systems. *International Journal of Communication Systems*, 33(15), p. e4547, 2020.

45. Gupta, R., A. Kumari, and S. Tanwar.: A taxonomy of blockchain envisioned edge-as-a-connected autonomous vehicles: Risk assessment and framework. *Transactions on Emerging Telecommunications Technologies*, pp. 1–23, 2020.

46. Kumari, A., R. Gupta, S. Tanwar, S. Tyagi, and N. Kumar.: When blockchain meets smart grid: Secure energy trading in demand response management. *IEEE Network*, 34(5), pp. 299–305, 2020.

47. Ge, Y., and G. Wei.: GA-based task scheduler for the cloud computing systems. *International Conference on Web Information Systems and Mining*. WISM, Sanya, pp. 181–186, 2010.

48. Ramezani, F., J. Lu, and F. K. Hussain.: Task-based system load balancing in cloud computing using particle swarm optimization. *International Journal of Parallel Programming*, 42(5), pp. 739–754, 2014.

49. Yu, Q., L. Chen, and B. Li.: Ant colony optimization applied to web service compositions in cloud computing. *Computers & Electrical Engineering*, 41(1), pp. 18–27, 2015.

50. Zhou, Y., L. Liu, L. Wang, N. Hui, X. Cui, J. Wu, Y. Peng, Y. Qi, and C. Xing.: Service-aware 6G: An intelligent and open network based on the convergence of communication, computing and caching. *Digital Communications and Networks*, 6(3), pp. 253–260, 2020, ISSN 2352–8648.

51. Al-Anbagi, I., M. Erol-Kantarci, and H. T. Mouftah.: A survey on cross-layer quality-of-service approaches in WSNs for delay and reliability-aware applications. *IEEE Communications Surveys & Tutorials*, 18(1), pp. 525–552, First-quarter 2016, doi: 10.1109/COMST.2014.2363950.

Chapter 7

Role of Blockchain Technology in Intelligent Transportation Systems

G. Kothai,[1] E. Poovammal,[2] and R. Durga[3]

CONTENTS

[1,2,3] Department of Computer Science and Engineering, SRM Institute of Science and Technology, Kattankulathur, India
Corresponding author: kg2247@srmist.edu.in

DOI: 10.1201/9781003264392-7

7.1 INTRODUCTION

The sixth generation integrates diverse technologies and applications; cate-nates everything; endorses holographic, space, haptic and underwater com-munication; and accommodates the Internet of Everything, the Internet of Nanothings, and the Internet of Bodies. The operating frequency of the sixth generation is 1 THZ, and for the fifth generation, it is 3–300 GHZ. Both the uplink and downlink data rates of 6G are 1 Tbps, and for 5G it is 10 Gbps and 20 Gbps. The operating frequency and data rates of 6G are higher than 5G's. The spectral efficiency of 5G is 10 bps/Hz/m^2, and for 6G it is 1000 bps/Hz/m^2, while the maximum mobility of 5G and 6G is 500 km/h and 1000 km/h, respectively. The spectral efficiency of 5G is lower than 6G's, and the maximum mobility of 6G is twice as high as 5G's [1].

The processing delay of 5G is 100 ns, and for 6G it is 10 ns. The traffic capacity of 5G and 6G is 10 Mbps/m^2 and 1–10 Gbps/m^2, respectively. The uniform user experience of 5G is 50 Mbps 2D, and for 6G it is 10G bps 3D. 6G is fully integrated with satellite, whereas 5G has no satellite integration. 5G is partially amalgamated, and 6G is fully amalgamated with AI, XR, Haptic communication, and automation. 6G accommodates more technologies and applications by prorating higher data rates, progressive reliability, low latency, and bulwark efficient transmission. The terahertz band, ranging from 0.1 THZ to 10 THZ, will play a pivotal role in 6G by prorated higher bandwidth, altitudinous capacity, ultrahigh data rates, and bulwark transmission [1]. This section furnishes an overview of intelligent transportation system (ITS), vehicular ad hoc network (VANET), and blockchain technology.

7.1.1 Blockchain Technology

An emerging technology that is decentralized and distributed ledger for all events and guarantees high security levels in all events [2] in different applications is blockchain technology. It is a secure and distributed structure with a transparent system that provides easy access to data for all users in the block [3]. This technology enables any transactions among authorities without requiring any entrusted third party [4]. It relies on validators who can supersede the third parties and endorse every transaction in a decentralized manner [4].

The chain constantly grows, wherein every new block is supplemented at the end, and the new block has allusions like hash value for the indication of the antecedent block [4]. The public and private keys are the two keys available for each node in the network. A Public key encrypts the messages that are dispatched to a node, and a proper private key decrypts the messages and permits a node to interpret the messages. This is called a public key encryption mechanism that assures consistency of messages, irreversibility of transactions, and nonreputability of a blockchain.

A cryptographic one-way hash function generates hash values for all the blocks in a blockchain and guarantees the immutability of messages, anonymity, and compactness of the block. The digital signature uses a private key for every transaction to provide authentication and morality of a transaction. Whenever a transaction is disseminated in a network, it is examined to be authentic. The network crowds the transaction into blocks that are timestamped by distinct nodes. Miners employ concrete consensus mechanisms like proof of work or proof of stake [4]. It is a P2P network; a conveyance is examined before it gets affiliated and may commune with all other nodes in a particular network [4].

The imperative tasks of nodes are liaison in the blockchain network, accumulating an updated ledger, validating and monitoring the transaction, passing on valid transactions, sensing for newly sealed blocks. Some of the essential intents of blockchains are monitoring digital assets like bitcoin and operating certain logic like smart contracts [4].

The three types of blockchains are only-cryptocurrency blockchain (C2C), cryptocurrency-to-business blockchain (C2B), and business-to-business blockchain (B2B) [5]. The only-cryptocurrency blockchain type deals with only cryptocurrency chains and is reserved for money or payment in decentralization. The mode of peer participation for the only-cryptocurrency blockchain is permission-less blockchain and public blockchain. Proof of work is a consensus mechanism, and it has high node scalability and low performance scalability. A built-in cryptocurrency called bitcoin is the cryptocurrency used for this type of blockchain. The bitcoin blockchain development addresses the issue of double digital money spending. The programming language used in C2C is C++ with very limited programming possibility, and the application development is very difficult [5].

The cryptocurrency-to-business blockchain is among the types of blockchain with a logic tier in the ledger and archives a multiaspirations programmable infrastructure. The public ledger depots the financial transactions and has amenities to marshal; executable programs on the blockchain are known as smart contracts. The small computing program that executes automatically when certain conditions are met is referred to as smart contracts. The cost for the verification, execution, and prevention of fraud is reduced by having tamperproof smart contracts. The permission-less blockchain and public or private blockchain comprise a mode of peer participation in C2B. Built-in cryptocurrency, called ether, is a cryptocurrency used. The programmable languages for C2B are smart contracts written in tandem with a high possibility for development [5].

The next blockchain that supports software execution for business logic without currency, such as hyperledger, is known as a business-to-business blockchain. The personalized application with respective needs must be served distinctively to each industry or business. This type of blockchain allows users to develop personalized private blockchain applications. It includes several sectors like finance, supply chain, Internet of Things, banking, manufacturing, and technology. A system that does not require the embedding of cryptocurrency or any form of mining operation is called a hyperledger [5].

The permissioned blockchain and private blockchain are the modes of peer participation in the B2B blockchain. The pluggable consensus algorithm is the consensus mechanism, and it has low node scalability and high performance scalability. No built-in cryptocurrency can be used, and chain code written in Golang for development at a higher level is the programmable language for B2B blockchain [5]. The public blockchain network, private blockchain network, and consortium blockchain network are the three kinds of blockchain networks [3].

The public blockchain network is also known as an open blockchain system where users do not demand permission from any authorities. Every process in the system can be supervised transparently. The private blockchain network also known as a special blockchain is wholly authorized by an authority. It permits participants in the network, operations in extracting data, and new transactions. The next is the consortium blockchain, where participation in the network is based on an approval mechanism and not open to everyone in the network. The system executes in a closed approach, and job illustrations are defined as unnecessary for reconciliation methods [3].

The several features of blockchain protocols are (1) an immutable, irreversible, distributed system, (2) no Centralized Authority, and (3) resilience [2]. Some of the key terms in blockchain technology are hash functions, node, cryptology, transaction, Merkle, tree, consensus protocol, P2P network, ledger, wallet, and blocks. Governance, health, finance, and technology are the various sectors of blockchain technology. Some of the application areas of the government sector are voting, land transaction, tax regulations, education, charities, labor rights, and authentication [3].

7.1.2 Intelligent Transportation System

An intelligent transportation system (ITS) is a future technology that helps in improving arterial safety, traffic management, and automobilist accommodation. It manages city-level traffic by providing efficient warning notification amid emergencies that may help in congestion control. It also includes providing cooperative awareness, giving emergency warning notifications, changing lanes safely, providing optimal traffic signal control, safe intersection crossing between lanes, allocations for parking, and downloading multimedia from the Internet. Providing security for various applications is a key challenge in the ITS [6].

ITS inaugurates a wireless affinity among conveyance, people, networks, and routes. Information is imperturbable by manifold resources on the network, and that information is processed, calculated, shared, and securely published for the people. Efficient traffic solutions are obtained by amassing, apportioning, and examining the information on route conditions, people flow, and conveyance [7]. Each conveyance in an intelligent transportation system can have active roles in ensuring safety and the free flow of traffic. The ITS roles can be as a sender for sending messages to the network, as a receiver to take messages from the network, and as a router to broadcast the information within the vehicular network [8]. Various standards are provided to intelligent transportation systems, and they can be interpreted in the following aspects.

1. The behavior of the product
2. Interface of devices
3. Auditing performance of system
4. Coordination and interaction of data
5. Product vendors, manufacturers, and the government

The first aspect is product behavior, which prescribes the ways that a product should behave. It helps in the understanding of a device and ensures uniform product behavior and product responses. The next is an interface that connects many devices with one another; i.e., manifold traffic signals should be channeled to the tantamount controller so that they work everywhere. Checking on the performance of a device is required because detecting for underperformance is imperative to prevent a device from overall impeding the system. Such evaluations enable manufacturers to develop cost-effective product that can set quality thresholds. The coordination and intersections help various agencies to transfer and store data effectively and in a standard format. Vendors select the best product for users, and the government provides requirements to the manufacturers in order to achieve uniformity for the product and its output [9].

The various application areas of intelligent transportation systems are classified into the following classes depending on the interfaces:

1. Center–roadside interface for transportation management
2. Center–center interface for communication centers
3. Center–vehicle interface for vehicle management
4. Roadside–vehicle interface for increasing the services of system
5. Roadside–roadside interface for interactions

A class can have subclasses or subgroups that use subsets of standards to be effective, and each subgroup may have more than one paradigm to pursue. The center–roadside interface standard exists between the center device and a roadside device. The communication takes place amid the transportation management center and roadway equipment [9]. The various fields of the center–roadside interface are as follows:

1. Collection and monitoring of data
2. Potent messages for system
3. Ramp metering
4. Traffic signals
5. Sensors for vehicles

The center–center Interface endeavors to make criteria for dissemination between management centers [9]. The various fields of the center–center interface are as follows:

1. Data archival
2. Traffic management
3. Traveler information

The center–vehicle interface provides criteria for dissemination between the management center and vehicle [9]. The fields of the center–vehicle interface are as follows:

1. Mayday
2. Transit vehicle communications

The roadside–vehicle interface is the interface that provides requirements for wireless communication amid roadside equipment and conveyance to enhance the performance and quality of the system [9]. The fields of roadside–vehicle interfaces are as follows:

1. Toll/fee collection
2. Signal priority

The roadside–roadside interface entails a principal for communication between roadside and railroad path accommodations, i.e., interaction between the road and rail equipment. Dynamic message sign standards and standard testing are also needed for intelligent transportation systems. Legislation for dynamic message signs can be employed by the dynamic message standard, which includes sign configuration, font configuration, sign control objects, message parameters, illumination objects, status objects, and power status objects. Standard testing helps in the evaluation of whether or not the standards made are effective. It is an essential part of any requirement [9]. It may be carried out using:

1. Validation testing
2. Verification testing
3. Experienced-based testing

Various types of an evaluation are required at different stages to minimize the risk of project failures [9]. The evaluation stages in intelligent transportation systems are:

1. Planning level stage
2. Deployment stage

3. Assessment
4. RP and SP survey

Some evaluation tools that help evaluate intelligent transportation system technology are traffic simulation models and ITS deployment analysis systems [9]. The safety and security threats of intelligent transportation systems are from humans, cyberspace, and physical systems. Driver behaviors, like inebriated driving, frantic driving, somnolent driving, and terrorist attack or misusing conveyances as weapons, will cause car collisions. Threat points in cyberphysical systems are pertinent conveyances, V2X security, and wireless implantable medical devices.

The vulnerabilities in cybersecurity for connected conveyances are a threat to communication links, to conveyance hardware like sensors and actuators, and to operating systems. The threats of V2X communication are to authentication, to message confidentiality and integrity, to bandwidth, and to the ITS network and component segmentation. The threat to implanted medical devices (IMDs) is to the battery, to the IMD firmware, and to communication links [10]. The mobility of nodes steers us toward the perpetual reworking of and dynamicity in node connectivity that incurs an idiomatic remonstrance to a blockchain system. Proof of work (PoW) attainment becomes hellacious due to the failing of nodes within the bounded length of time designated to barter a new block for verification. The probability of annexing blocks to the chain, the stability of a milieu, and the enumeration of blocks in a milieu are the three key metrics in blockchain [11].

7.1.3 Vehicular Ad Hoc Network

Intelligent vehicular ad hoc network (VANET) are the main elements in ITS that have gained much attention from the government, academic institutions, and industry [12]. VANET is an upcoming technology that provides security and traffic information to the passengers by providing salable and infotainment information to automobilists and passengers [13]. Safety and nonsafety are the common categories of the VANET application. The safety application includes collision warning. The nonsafety application includes providing real-time traffic congestion reporting to detect traffic jams, gain routing information for traffic management, ensure high-speed tolling for conveyances, and provide mobile infotainment for the users [12].

The two branches in intelligent VANET are vehicle-to-vehicle (V2V) communication and vehicle-to-infrastructure (V2I) communication, as delineated in Figure 7.1. V2I can be called vehicle-to-roadside (V2R) communication [8, 14]. V2V communication takes place between vehicles in the network. It includes an emergency braking system that facilitates drivers to anticipate potential mishaps. V2V shares data among conveyances to alert drivers to

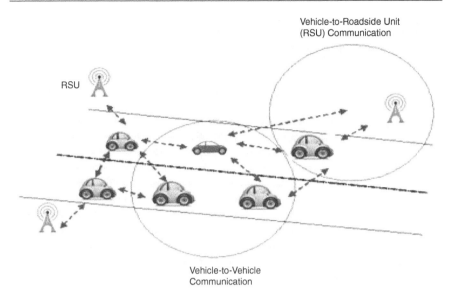

Figure 7.1 Vehicular communication.

precarious circumstances and to provide warning signals to drivers before a collision occurs. Infotainment, such as a braking system, is shared among the conveyances in the network through a dedicated short-range communication (DSRC) protocol that warns of collision conditions, reducing fatal crashes between conveyances and enhancing roadway safety [14].

Naïve broadcasting and intelligent broadcasting are the two kinds of message forwarding techniques in intervehicle communication. When conveyances in the network broadcast messages periodically at regular intervals, that is naïve broadcasting. The conveyance ignores the message based on the sender location: If the message is broadcast from a conveyance behind the receiving conveyance, the message is ignored; if the message is from conveyance ahead of the receiver, then the acquiring vehicle disseminates the message to conveyances behind it. The drawback faced by this type of broadcasting is that, when large amounts of messages are broadcast, a collision is possible, leading to lowered communication in the form of lower message delivery rates [8].

Intelligent broadcasting, depending on acknowledgments, limits the messages to be broadcast for any emergency occurrences that address the essential problems of naïve broadcasting. In intelligent broadcasting, if a conveyance acquires the message from behind it, then at least one conveyance has received the message, and it disseminates. When a node obtains a message from more origins, it reacts to the first received message [8].

Some of the advantages of technologies in vehicle-to-vehicle communication are a collision avoidance system for traffic management, a driver warning system for safe lane changing, automated emergency braking assistance to avoid accidents, and a vehicle permanence system for users [14].

When the dissemination occurs among conveyances and the roadside unit in the network, it is vehicle-to-infrastructure communication. This plays an intrinsic role in gathering information on traffic and roadway circumstances for suggesting node behavior. Through the dedicated short-range communication (DSRC) protocol, the exchange of safety and operational data among conveyances and roadside units (RSUs) alleviates the probability of conveyance accidents [14].

The configuration of V2R communication provides higher bandwidth links among conveyances and RSUs. The roadside unit determines the proper speed according to its internal schedule and traffic situation by broadcasting dynamic speed limits. The message contains information, such as speed limit, and the roadside unit broadcasts so as to identify any conveyances requiring a speed limit warning in the immediate area. A broadcast in the listening area is made to a driver to reduce speed when the conveyance is violating the appropriate speed limit [8].

Some of the safety applications of vehicle-to-infrastructure communications are providing messages to conveyance operators and possibly to authorities of red light violations, violating speed zone limits, on-the-spot weather information, stop sign violations, illegal railroad crossings, and oversize conveyances [14].

The trusted authority (TA) of the system, the onboard unit (OBU) of a conveyance, and the roadside unit (RSU) of the network are the three major components of the system model in VANET. The communication patterns of vehicular communications in VANET can be divided into four categories: (1) warning message propagation among conveyances, (2) group communication among conveyances, (3) beaconing among conveyances, and (4) warnings between infrastructure and conveyances [15].

The communication media, or paradigms, for VANET encompass dedicated short-range communication protocol and the degree of IEEE 802.11 and wireless access in the vehicular environment (WAVE). The principal attributes of VANET are (1) integrity, (2) confidentiality, (3) authenticity, (4) availability, and (5) nonrepudiation [15]. The important requirement and current issue faced by vehicular ad hoc network is security. Several security disputes with regard to VANET have to do with such features as authentication, availability in the network, integrity of a message, confidentiality, and nonrepudiation of the message.

Security attacks are broadly categorized into the five features of vehicular ad hoc networks [15]. The unique characteristics of VANET are mobility, dynamic network topology, real-time constraints, computing and storage capability, and volatility. Mobility describes the moving of conveyances in VANET at high speed. Due to the conveyances' high mobility, there is a quick change in topology that makes the VANET vulnerable to assorted attacks like malevolent conveyances.

Real-time curtailment explains the imparting of information within a VANET-designated time range that gives the receiver sufficient time to execute appropriate reactions. Processing a greater amount of information among nodes and RSU is a challenging issue in VANET in terms of computing and storage capabilities. Volatility describes the dynamic connection between two nodes in VANET that would continue to exist for a few wireless hops. Anticipating the difficulties with securing personal contacts in VANET also comes under these characteristics [15].

The various application services of VANET are providing conveyance safety, enhancing navigation, assuring automated toll payment, ensuring traffic management, and providing services based on locations such as detecting fuel stations, hotels, or restaurants in the immediate area and infotaining the applications via Internet access [8]. Blockchain assures a high level of security for all events with its decentralized and distributed ledger features in the diverse applications of VANET [2].

7.1.4 Motivation and Scope of Blockchain in 6G for Intelligent Conveyance

Blockchain is reckoned as a key enabler for autonomous conveyance communication in the 6G network. Trajectory concerting, augmenting air traffic, and abundance in contingency can be abated by this distributed technology [16]. Blockchain steps up privacy and security for autonomous conveyance with its immutability, reliability, transparency, and traceability [17], in addition to its adaptability and scalability. Blockchain furnishes a decentralized environment for autonomous conveyance, and 6G technology heightens the reliability of connectivity [16].

7.1.5 Contribution of Blockchain in Intelligent Conveyance

The upshot of blockchain technology in vehicular ad hoc networks is that it can enumerate the nodes propagating in a block of VANET, which is complex due to the nodes' mobility. It is arduous to communicate the data between nodes without any central infrastructure. Forming a space for blockchain nodes and the continuous stirring of nodes is a hellacious task in VANET [11]. A single point of failure becomes possible due to the dependence on a trusted central entity, which is needed in any centralized approach to the security of VANET [18].

Blockchain technology in VANET stores data in a chain, cinching that data is tamper resistant and can be accessed using cryptography and hash functions in the blockchain. Consensus protocol can be used to overcome the data viscidity that can be problematic in untangling the webs of security and to replace the mainstay of a trusted central entity in the VANET environment [18].

7.2 SYSTEM MODEL FOR VEHICULAR AD HOC NETWORK

The sixth generation enables cell-free communication that can be fully customized for unmanned aerial vehicles (UAVs). The doper accouterment[1] is catenated to the whole network, not to any specific cell that consummates the knot of cell coverage. It furnishes the best coverage without any special configuration on the device [1]. The three important modules of the system model in VANET are the trusted authority (TA) of a system, the onboard unit (OBU) in a conveyance, and the roadside unit (RSU) of the network. The other components in VANET are speed-sensing nodes (SS), aggregator fog devices (AFD), and conveyances [19].

7.2.1 Trusted Authority

Trusted authority (TA) is the entrusted authority center in VANET, similar to a cloud server that administers the rostrum of registration and verification for OBUs and RSUs [19]. It is also accountable for the intended security management, which includes revoking the nodes when a conveyance broadcasts fake messages or exhibits malevolent behavior and verifying genuine conveyances. The obligations of TA are very high computational power and ample storage capacity [15].

7.2.2 Onboard Unit

The onboard unit is a processing unit implanted in a conveyance that collects information from an assortment of sensors [19]. An OBU, when it is equipped in a conveyance, employs a transceiver to convey messages with other onboard units and roadside units [15]. Sensors like a global positioning system (GPS) helps in collecting the information that must be dispatched to the OBU. The OBU also observes and accumulates the information to manifest messages consigned to neighboring conveyances and roadside units via a wireless medium [15]. It includes current information such as the conveyance position, acceleration carried out in a conveyance, driving directions for lane changes, and traffic information for traffic management on the roadway [19].

7.2.3 Roadside Units

A roadside unit is an essential infrastructure contrived at designated locations like intersections and parking lots in VANET [15, 19]. Among the tasks of roadside units is to broaden the communication range for imparting information to other OBUs and RSUs in running safety applications [15].

7.2.4 Speed-Sensing Nodes

The speed sensing nodes are built-in speed sensors that dispatch reports for violating traffic by a conveyance over a specified time. These sensors are also miniature devices such as smartphones, and the sensing nodes are capable of dispatching the traffic violate report (TVR) to the aggregator fog devices [20].

7.2.5 Aggregator Fog Devices

The aggregator fog devices (AFD) are lightweight servers whose devices, fixed along the routes, have computation, storage, and proclamation capabilities. The aggregator fog devices are connected with speed-sensing nodes wirelessly. Some of their tasks are to collect traffic violation reports (TVR), aggregate those reports, and effectuate signature verification on agglomerated traffic violation reports. Excess speed warning messages from traffic violation reports are broadcast to all authorized entities and forwarded to other aggregator fog devices and roadside units [20].

7.2.6 Vehicle

All the intelligent conveyances in the network must be fitted with speed sensors and affiliated with a global positioning system (GPS) to check the speed limit of a conveyance in every region. The onboard unit amasses the data, such as speed from speed sensors, and sends it to aggregate fog devices. All the conveyances in the network must be registered with the trusted authority and the traditional identifier of the conveyance [20].

The system model consists of security requirements and preliminary knowledge. The basic security requirements for recording offenses in vehicular ad hoc networks of security and privacy are identity privacy preservation, rapid vehicle revocation, nonrepudiation, traceability, unlink capability, and modification of passwords. In identity privacy preservation, the offending conveyances cannot acquire the true identity of the node; only the TA can pursue the authentic identity [20].

Traceability is the ability to track the authentic identity of the offending conveyance based on its data and to assent to appropriate action when the entrusted organizations can promptly detect the valid identity of the conveyance and cease that conveyance from persisting in disseminating messages.

This is known as rapid vehicle revocation. Nonrepudiation is the inability of the conveyance to deny the sent message after sending it [20].

When the offending conveyance or an attacker cannot find the messages of the same conveyance based on the information included in these messages, that is known as unlink ability. The operator who feels that the contemporary password is not protected can amend the password offline, relying on the requirement of password modification to improve the security and feasibility in VANET [20].

Some preliminary pieces of information required for system models in vehicular ad hoc networks are secured one-way hash function, elliptic curve discrete logarithm problem (ECDLP), computation Diffie–Hellman problem (CDH), and Bloom filter [20]. The messages are transmitted between on-board unit and roadside unit through the wireless channel. Information between the trusted authority and roadside units can also be transmitted via wired channels. The transmission of information between onboard units on different conveyances is also carried out in wired channels [20].

7.3 FEATURES OF THE VEHICULAR NETWORK AND COMMUNICATION MEDIA

Fifth generation apportions its standard infrastructure for accommodating an assortment of technologies, such as artificial intelligence, mobile broadband communication, Internet of Things, self-driving cars, and smart cities [21]. This section discusses the main features of vehicular networks and communication standards for VANET.

7.3.1 Factors of Vehicular Network

Most of the conventional ceiling can be a maneuver by blockchain and distributed ledger technology, and it also expedites the functional standards of 6G [21]. The security requirements needed for VANET are availability of resources, integrity of the messages, authenticity of the conveyances, nonrepudiation of nodes, and confidentiality between users [15].

7.3.1.1 Availability of Resources

In safety applications for safety messaging, the wireless channel must be working so that the conveyances in the network can dispatch and acquire the warning messages [4]. Having in addition to deal with malicious or faulty nodes, availability ensures that all the normal operations are carried out in a network and its applications [15]. Some of the kinds of attacks on availability are denial of service (DOS) attack, jamming attack, broadcast malware attack, black hole and gray hole attack, tampering attack, greedy behavior attack, and spamming attack [15].

7.3.1.2 Confidentiality between Users

Data must be accessed only by the designated receiver, and no other nodes in the network must be privy to the confidential information sent from the sender [15]. It must also be guaranteed that the data dispatched from the sender to the beneficiary is not eavesdropped [12]. An eavesdropping attack and a traffic analysis attack are the two types of attacks on confidentiality in VANET [15].

7.3.1.3 Authenticity of Vehicles

The important requirement for vehicular ad hoc networks is authenticity [12], which denies malicious entities' entry and acts as a defense against various attacks in the vehicular ad hoc network [15]. Authentication is composed of two classes: ID authentication and entity authentication. When the message to the receiver from the sender is correctly identified and verified with a unique ID, that is known as ID authentication. The ID of the conveyance is the vehicle registration or license plate in VANET. Entity authentication assures that the lately obtained message is new and ascertains that the message is emitted and acquired promptly [12]. The various kinds of attacks in authenticity are Sybil attack, tunneling attack, GPS spoofing attack, and free-riding attack [15].

7.3.1.4 Integrity of the Messages

Ensuring that a message is not amended during data dissemination from sender to receiver is known as integrity. This also protects information against illegitimate users, specifically prohibited inception, demolition, or amendment of data [15]. The message suppression attack, masquerading attack, and replay attack are integrity attacks in the vehicular ad hoc network [15].

7.3.1.5 Nonrepudiation of Nodes

Nonrepudiation is when the sender and receiver cannot contradict their dispatch or acceptance of a message in any kind of situation [15]. It also ensures that any accidents caused by the driver must be reliably identified [12]. The repudiation attack is the only attack carried out against nonrepudiation in the vehicular ad hoc network [15].

Access control and privacy are two other requirements for the safe messaging of a security system in a vehicular ad hoc network [12].

7.3.1.6 Access Control

Access control discriminates between divergent accessing levels of a node or infrastructure constituent for an application. It also specifies the access level of each node in a network through some specific system-wide policies [12]. Access control also removes misbehaving nodes from the network by using systems that detect intruders and a scheme for trust management where the abrogation of certificates is done in VANET [12].

7.3.1.7 Privacy

Privacy is one of the important factors because the invasion of drivers' personal data can violate their privacy. Some information, such as trip details, time, vehicle identifier, location of the conveyance, technical description, is vulnerable and may cause inferences about drivers' actions and intentions to unauthorized observers [12]. It must also ensure that driver identity and conveyance location are not exposed [22].

7.3.2 Communication Media

Intelligent resource management, elevated security features, and scalability are some of the features of blockchain for the sixth generation [21]. The three main communication standards that deal with VANET are dedicated short-range communication (DSRC) protocol, degree of IEEE 802.11p, and wireless access in the vehicular environment (WAVE).

7.3.2.1 Dedicated Short-Range Communication

DSRC is a short to medium-length memorandum service for vehicular communications. It covers a hefty number of applications for communications that includes providing safety messages for intervehicular communications, collecting tolls, obtaining the information about traffic, offering the drive-through payment, and others. In smaller communication zones, the DSRC protocol aims to provide lower communication latency and higher data transfer [8]. The Federal Communication Commission (FCC) has allocated 75 MHZ of spectrum and band width from 5850 to 5925 GHZ for DSRC. Some service channels in DSRC are for critical safety and public safety [15].

7.3.2.2 Wireless Access in Vehicular Environment

The set of stereotyped protocols, utilities, and related functions used to constitute vehicular communication are defined by WAVE [15]. The challenges faced by various traffic scenarios in VANET are varying the speeds of the conveyance, providing the variations in traffic patterns, and driving in different settings.

Some of the problems faced by the traditional media access control (MAC) are dealing with the major overheads in vehicular scenarios, the need to ensure safety communications between the conveyances, and the fast exchange of data. To inaugurate communication, the examination of channels for beacons with multiple handshakes from any access point is needed and is associated with heightened complexities and overheads. The challenging requirements of IEEE media access control operations are addressed by WAVE in the sixth generation by furnishing higher frequency in blockchain armature [8].

7.3.2.3 IEEE 802.11p

The degree of change in IEEE 802.11p from the traditional IEEE 802.11 protocol accommodates vehicular networks and designates the characteristics of the physical and medium access layers for VANET [8].

The three basic communication patterns in secure communication protocols for VANET are beaconing, restricted flooding, or geocasting, and geographic unicast routing [12].

7.3.2.4 Beaconing of Messages

Beaconing is a mechanism that divulges information through a single hop that is not imparted by the acquiring nodes in VANET. Some of the information, like the conveyance's own position, speed, or heading, is included in secure beaconing. The authenticity and integrity of the beacon messages are verified based on the following:

1. The sender must be a legitimate user of the network.
2. The data or information must be maintained regularly.
3. Alteration of the data is not possible.
4. The current vehicle identifier and its location must be included.

The challenge faced by secure beaconing is the higher frequency of the applications that can be granted by the blockchain armature [12].

7.3.2.5 Secure Restricted Flooding/Geocasting

Flooding disseminates information among the arbitrated surroundings of a conveyance for various applications in VANET. Multihop broadcast forwarding is the principle by which every node rebroadcasts the messages. The rebroadcast messages are restricted by time-to-live (TTL) or a geographic destination area (GDA). Preventing the malicious conveyance in rerouting the messages, tampering with the data, or dropping the packets; addressing attacks that aim to rattle the entire operating network; and ensuring integrity, authenticity, and reliability are the goals of secure restricted flooding/geocasting [12]. The data about spectrum is immutably recorded in the blocks. Recording the data along with the blockchain is the purpose in spectrum governance. The blockchain armature excludes fraudulent users based on the transparent behavior [23].

7.3.2.6 Secure Geographic Routing

The multihop single path forwarding technique is a principle of geographic routing where data destination is a coordinate contrary to a node address. This forwarding approach helps in passing the messages along to the node geographically closer to the recipient node than the current node is. Every node requires knowledge about its one-hop neighbors and their actual locations so that it can select the next hop for a data packet [12]. Integrity of the packets must be protected, and they must be guaranteed to arrive at the legal members of the network [12]. 6G technology integrates the applications of satellite types like telecommunication multimedia network, navigation, and weather information services; combines artificial intelligence and nanocore; and handles the abundance of data and very high data rate connectivity per device.

The heterogeneous network and D2D communications are two network architectures intended by the 5G fog-based Internet of Conveyances [20].

7.3.2.7 Heterogeneous Networks

Heterogeneous networks aim to attain higher data transmission rates and higher network capacity in the network. These two characteristics of heterogeneous networks increase spectral efficiency through smaller cells. They offer high data rates that operate within the realm of 30–300 GHZ and 1–10 mm for the spectrum and wavelengths using the millimeter wave spectrum [20].

7.3.2.8 D2D Communications

Enabling the speed sensors to interact with particular fog devices within the authorized cellular bandwidth without examining the fixed stations is known as D2D communications. This furnishes the memorandum between conveyance, speed sensors, and fog devices and is also known as millimeter wave technology [20].

The design goals of the communication model are security and performance. The security objectives in the design goals of the communication model in the vehicular ad hoc network are privacy preservation, mutual authentication, data confidentiality, integrity, authorization, key escrow resilience, and traceability. Communication and verification overhead, robustness, and lightweight are the performance objectives, which involve the sender, receiver, speed sensing node, aggregator fog devices, and traffic violation report [20].

7.4 SECURITY ATTACKS IN VEHICULAR COMMUNICATION

The drawback of centralized autonomous conveyance, like single point delinquency, is that it could prostrate the entire network. However, ceiling restraints on data, low delineation and accountability, and circumscribed scalability can be alleviated using blockchain technology. Blockchain, hinging on an autonomous conveyance system, enables both automobilist and passengers to be safe, secure, and comfortable [17].

Attackers possess various capabilities and intentions for malicious behavior in order to break into system facilities and either to gain access to its confidential data or to disrupt the functionality of the network [24, 25]. The classification of these attackers, delineated in Figure 7.2, based on:

1. Membership
2. Activity
3. Intentions

Based on Membership

When any malicious activity is carried out in a network by any authorized or unauthorized member node, the membership function highly affects the impact of the attacks and its prevention. This basis is divided into two types of attackers [25]:

1. Internal Attackers
2. External Attackers

When the malicious activity or behavior is perpetrated by the authorized member node in the network, either to get personal gain or to agitate the network's

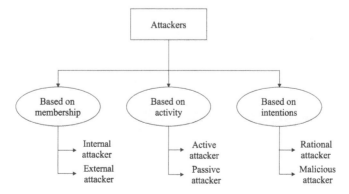

Figure 7.2 Types of Attacker [25]

functionalities, the attackers are said to be insiders or internal [25]. External attackers are interposers who penetrate the network and impersonate an authorized constituent node in order to perform the malicious activity within the network [25].

Based on Activity

Depending on the attacker's activeness and frequency of changes to the network, attackers based on activity are classified into two types [25]:

1. Active attackers
2. Passive attackers

Generating malicious packets and signals or altering information in the network are the activities of active attackers. The attacks caused by active attackers are more impactful than passive attackers [25]. Passive attackers do not alter any information in the network, but they sense the whole network silently [25].

Based on Intentions

If attackers are associated with any intention behind their attacks, then is the attacks are based on intentions. There are two types of attackers based on Intentions [25]:

1. Rational attackers
2. Malicious attackers

Rational attackers are more predictable as they seek personal benefit from the attacks on the network [25]. Malicious attackers will create any obstacle to disrupt network functionality, but they will not gain any personal benefit from the attacks [25].

Intentional attacks are divided into five types, as shown in Figure 7.3. These are attacks on:

1. Availability
2. Confidentiality
3. Authenticity
4. Integrity
5. Nonrepudiation

7.4.1 Attacks on Availability

All participating members of the blockchain network corroborate a transaction before it goes on to the block. The blockchain fortifies fence communication by immutability, reliability, transparency, and traceability [26]. The

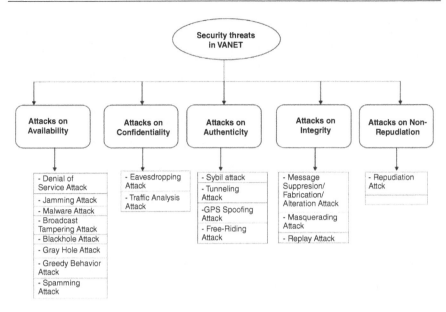

Figure 7.3 Security threats in VANET [15].

following are the various attacks that come under availability: denial of service attack, jamming attack, malware attack, greedy behavior attack, black hole and gray hole attack, broadcast tampering attack, and spamming attack.

7.4.1.1 Denial of Service Attack

The most ubiquitous meddlesome attack against availability is the denial of service (DoS) attack [22]. The attackers will perform a DoS attack by obstructing the communication media or channel or network. Overriding the resources in VANET may make it impossible or complicated for legitimate users to obtain any information and resources in the network [15, 27]. Cryptographic solutions, IP-CHOCK, anticolony optimization, frequency-hop or channel switching, and the preauthentication technique provide solutions for the DoS attack in VANET [24, 15].

Cryptographic solutions include Time Efficient Stream Loss Tolerant Authentication (TESLA), TESLA++, Elliptic Curve Digital Signature Algorithm (ECDSA), VANET Authentication using signatures and TESLA ++ (VAST), prediction-based authentication (PBA), and Fast Authentication (FastAuth) and Selective Authentication (SelAuth) [28]. When the attackers attack in a distributed manner from different locations aiming to impede authorized users from ingress, the information or resources in a network are said to be under a distributed denial of service (DDOS) attack [24, 15].

7.4.1.2 Jamming Attack

The jamming attack is a serious security threat in VANET [24]. Disrupting the communication channel by the attacker to exploit the persistence signal with an analogous frequency is a jamming attack [15]. When the signals are sent repeatedly by the jammers, the victim feels that the state of the channel is still busy, so they cannot send or receive packets in that area. In a jamming attack, the sender can send the packet successfully, but the receiver will not be able to receive all the packets sent by the sender. This may lead to a low packet delivery ratio, and when the receiver does not receive adequate information like road conditions, weather, or accidents, that may cause fatalities [24].

7.4.1.3 Malware Attack

The attackers penetrate the vehicular ad hoc network by installing malware like viruses into the OBUs and RSUs to destroy the normal functionalities of the network [15]. Very often, malicious insiders execute this type of malware attack rather than outsiders. Normally, VANET is affected when performing a software update for OBUs in conveyances and RSUs. In the VANET research community, embedded antimalware frameworks are still considered problematic [24].

7.4.1.4 Broadcast Tampering Attack

Preventing successful safety messages to authorized nodes or conveyances in the network by an internal attacker that broadcasts fake warning messages is known as a broadcast tampering attack [15]: hidden vehicle attack, GPS spoofing, tunneling, and location tracking. Black hole attack, wormhole attack, gray hole attack, and rushing attack are the classes of attacks for tampering with routing information. Illusion attacks and greedy behavior attacks are the two classes of attacks in tampering with technology. Some of the classes of attacks for tampering with identity are Sybil attack, masquerading, impersonation, ID disclosure, key/certificate replication, brute force attack, and repudiation attack. Bogus information, bush telegraph, message tampering, message fabrication, message suppression, broadcast tampering, and spam attack are some of the classes of attacks for tampering [29].

7.4.1.5 Black Hole Attack

Among the several network threats, one is redirecting the network traffic by a malicious node [24] or when the informant node or a neighboring conveyance sends the route request (RREQ) packet to a malicious node within the network and then responds quickly in replying to the route reply (RREP) packet to that particular node. In most scenarios, that malicious node will be the first node to respond to the route request, and it will not even check the routing table before sending the route reply packets. The malicious node drops all the data packets received that are consigned by the route request cradle node. The preliminary stage and secure stage are the two stages that help detect the black hole attack in the network [30]. Detecting the black hole based on cross-layer includes cooperative watchdog, MAC monitoring, and cross-layer design [31].

7.4.1.6 Gray Hole Attack

In a gray hole attack, the malicious node drops the data packets sent from sender nodes in the network, and all other control and routing packets sent from other nodes are forwarded to the compromised or attacked node. After dropping the packet, that malicious node will act like a normal node; hence it is difficult to identify the malevolent node in the VANET [32]. The ABC algorithm and the Genetic algorithm are the two approaches that mitigate the effects of the gray hole attack [33].

7.4.1.7 Greedy Behavior Attack

When the malevolent conveyance increases the bandwidth at the expense of any other nodes in the network in order to maltreat the MAC protocol in VANET, that is said to be a greedy behavior attack [15]. The dubiety phase and the deliverance phase are the two phases for detection algorithms to identify greedy behavior attacks in VANET. This detection algorithm combines both the linear regression and the watchdog notions for VANET. The suspicion phase includes the correlation coefficient, linear regression, and the watchdog supervision tool. Inputs, fuzzy set and membership function, fuzzification and membership degree, decision rules, and defuzzification are the steps involved in the decision phase of the vehicular ad hoc network [34].

7.4.1.8 Spamming Attack

When an abundance of spam messages are injected into the VANET system by the attacker, the messages will reside in the network's bandwidth, which may lead to collisions in the vehicular ad hoc network [15].

7.4.2 Attacks on Confidentiality

Blockchain technology hoards all the transactions effectuated by autonomous conveyance and roadside infrastructure in the blocks of the network [17]. Eavesdropping attacks and traffic analysis attacks are the two type of attacks involved in confidentiality.

7.4.2.1 Eavesdropping Attack

When the attacker extracts any confidential information from the preserved data, such as stealing the identity of a conveyance or tracking it by collecting location data from the network in the vehicular ad hoc network is known as an eavesdropping attack [15].

7.4.2.2 Traffic Analysis Attack

When messages are transmitted between OBU and RSU, the attacker in the network listens and analyzes their frequency and duration for purposes of collecting confidential information in the vehicular ad hoc network, this is called a traffic analysis attack [15]. The plausibility validation network (PVN) and PVN paradigm on traffic safety applications are two defenses against this attack. The PVN paradigm on the traffic safety application provides the following rules:

> *Rule 1*: The replicate message ought to be dropped.
> *Rule 2*: Depending on the nature of the event, the range to disseminate the information should be reasonable.
> *Rule 3*: The beneficiary should check the probability of the location, timestamp, and velocity of the messages [35].

7.4.3 Attacks on Authenticity

The sixth-generation technology advocates communication among UAV base stations and satellite, with the UAV network secured by blockchain technology [16]. The following are the several attacks that fall under authenticity.

7.4.3.1 Sybil Attack

The Sybil attack is one of the critical attacks in vehicular ad hoc networks [24], where multiple fake identities are created to fragment the everyday operations of the network. It also propagates erroneous information regarding traffic in the network to change the routes of authorized users and to clear the road [15, 36]. Spreading false messages, modifying the messages and replying, dropping the packet, forging GPS information, and modifying the transmission power are operations performed by attackers using Sybil identities in the VANET [36].

Some of the espial and hindrance mechanisms of Sybil attack are espials of Sybil attack using nearer nodes, the platoon dispersion detection technique, cueing by examining the behavior of the receipt packets, cueing by cryptography, timestamp series approach, parameters calculation and observation, RSS espial approach, and group leadership. Periodic communication, group construction of neighboring nodes, exchange groups with other nodes in the nearby area, and the identification of conveyances that comprise similar neighboring nodes are the four phases of overall Sybil node detection [36].

7.4.3.2 Tunneling Attack

The tunneling attack is a dangerous attack, analogous to the wormhole attack in VANET [15]. The two distant parts in VANET are connected via tunnel or auxiliary information channel by the attackers so that the significant-distance node can associate as a neighbor [15]. Due to the wireless nature of the network, a wormhole can be created by any malevolent node in the network. This wormhole attack creates a tunnel for two or more malevolent nodes where the data packets can be transmitted from one malevolent node to the other malevolent node, and then these packets are broadcast at the end of the tunnel [37].

The unauthorized access of resources, performing a DoS attack in the network, disrupting the routing process for message delays, busting the security of data packets during transmission are performed in wormhole attacks. The temporal leash and TESLA with instant key disclosure (TIK) protocol are the two approaches that help in preventing wormhole attacks. Shared key distribution with a public key can be used for secure message broadcasting of wormhole attacks in vehicular ad hoc networks [37].

7.4.3.3 GPS Spoofing

Attackers sometimes make other nodes in the network believe that the conveyance is available in different locations by generating a false global positioning system (GPS) that is different from the actual signals from the entrusted satellites [15]. Some techniques for preventing GPS spoofing are the hindrance of timestamp vault, short-lived pseudonym certificates, and retrospective attack detection via logging. The hindrance of the timestamp vault includes one-directional timestamp jumps and multiple independent time sources [38].

The multiple independent time sources available are local time sources, alternative wireless reference time sources, and noncontinuous updates from secure reference time sources [38]. The different forms of berth pacing attacks are pacing random berths using single ID (FRPSI) and multiple IDs (FRPMI), pacing path using single ID (FPSI), and multiple IDs (FPMI) [39].

7.4.3.4 Free-Riding Attack

In a free-riding attack, any selfish conveyance takes advantage of another's authentication contributions in a cooperative authentication scheme without making its own authentication. This may lead to a serious threat to cooperative message authentication [15]. The selfish behavior of the free-riding attack is generally divided into two categories: free-riding attacks without endorsement efforts and free-riding attacks with pretend endorsement efforts [40].

7.4.4 Attacks on Integrity

Blockchain technology is a doable solution to furnish security for autonomous conveyance that earns credence among autonomous conveyances, bulwarks data modification, brisk data access in distributed systems, and no single point delinquency [17]. The following are some of the attacks that have occurred based on integrity.

7.4.4.1 Message Suppression/Fabrication/Alteration Attack

In the suppression of messages or the fabrication of data or alteration in any information attack, the attackers alter some part of the information during its transmission for some unauthorized effect [15]. When any malicious node in the network creates false information and broadcasts it to all the neighboring nodes in the network to claim access to packets in the transmission, this is referred to as a message fabrication attack. The algorithms for detecting the message fabrication attack are acceptance-range-based verification, trusted-time-based verification, and node-availability-based verification [41]. The conveyance trust model for suppression attacks is the local entrusted conveyance model and TA entrusted vehicle model [42].

7.4.4.2 Masquerading Attack

This type of attacker steals the passwords of legitimate users and enters the vehicular ad hoc network as an authorized user in order to broadcast false messages [15]. The RSS property of wireless communication helps detect unauthorized, nonvalid, or spoofed identities [43].

7.4.4.3 Replay Attack

A replay attack is the attack where the attackers reinject the beacons and messages continuously received before in the network to confound the traffic authorities in discerning nodes in any contingencies [15].

7.4.5 Attacks on Repudiation

Security and privacy can be furnished by the hyperledger armature for crams in the UAV network [16]. The repudiation attack is the only attack based on the nonrepudiation service.

7.4.5.1 *Repudiation Attack*

If the assailant in the network contradicts the delivering or obtaining of any vital messages, leading to a dispute in the vehicular ad hoc network, this is called a repudiation attack [15].

Blockchain technology, with its elevated security features, unleashes the capacity of 6G systems in terms of authentication, integrity, confidentiality, and availability. The data confidentiality attack in the 6G network will be amended by the immutable and tamperproof cryptographic nature of blockchain—the effectiveness of the ultrahigh data rate in the 6G network's superiority over data integrity attack. The communication of information over the network in the blockchain armature will skunk any issues of data integrity. Security attacks will have such high connectivity in the 6G network that it will deny availability attack issues of VANET [44]. Storing the data in a decentralized distributed manner dodges the possibility of single-point failure in data authentication attacks. The traceability feature of the blockchain armature dodges the repudiation attacks in VANET [45].

7.5 SECURE VEHICULAR COMMUNICATION USING BLOCKCHAIN TECHNOLOGY

The basic features of blockchain technology are immutability, a distributed and trustless ambient, privacy and anonymity, blistering transaction speed, reliable and precise data, and transparency [46]. A trustworthy message propagation is an issue in VANET that can be solved using blockchain technology. Immutable distributed public databases can be used for bulwark message dissemination in the VANET so that any node in the network can vet the information [46]. This new type of blockchain bolsters the node trustworthiness and message trustworthiness in the distributed ledger in the vehicular ad hoc network for secure message dissemination that can be managed independently within the country [46].

Ethereum is a decentralized platform, and it aims to create contracts in blockchain. Unlike in the conventional VANET infrastructure, the RSU not only is used for the communication relay station but also grants the contingency of Ethereum-based applications that can be disseminated to the Ethereum blockchain [47].

This Ethereum blockchain is the combination of VANET that deploys the application for self-governed and decentralized network operation. Traffic regulation applications for traffic management and for tax and insurance applications for conveyances are the mandatory applications that can impose network rules. There is also a higher possibility of deploying other applications using Ethereum Turing complete high-level language [47].

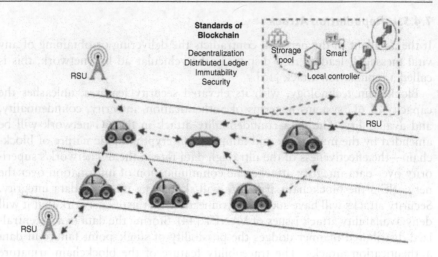

Figure 7.4 Blockchain technology in VANET.

Ethereum is generating a scheme that guarantees the reliability of the node and memorandums passing by confirming them in a blockchain in VANET for other automobiles. A public blockchain can be used for storing and managing all the conveyance nodes and message allegiance information in a geographical realm, as shown in Figure 7.4. The freshness of the messages is maintained by the blockchain technology using timestamps and hashing techniques delivered in VANET [48].

The principles for the architectural design are as follows:

Principle 1: The ASES (very large networks or groups of networks with a single routing policy) or third-party services provide the on-demand services for DDoS detection and mitigation [48].

Principle 2: The attack information must be received or reported as it is necessary to dedicate a conveyance node allied to the blockchain for the domain [48].

Principle 3: Coordinating the attack recoil, the DDoS appeasement modules must be activated on the blockchain entities, enumerate the IP addresses, and report to the blockchain efficiently [48].

Principle 4: Reporting the IP addresses to the smart contract can be done only by the patrons with the testimony of possession [48].

Principle 5: Different security policies in different domains with different indexing management systems must be implemented [48].

The customers, ASES, and blockchain or smart contract are the three components of architectural design [49]. The design also includes the device and ontogeny of architecture to repel DDoS attacks based on blockchain technology across different domains; adoption and integration are facilitated to impose criticality on the ASES by the SDN and to utilize additional security mechanisms without modifying the existing ones [49].

In the toll payment application system, the client must authorize the conveyance with the unique ID to distinguish the electric conveyance by driver. The participant attempts to exploit the hyperledger compiler and is lodged in the blockchain. Analogously, customers in the infrastructure are compounded to the network in VANET. If any transaction takes place using driver details and infrastructure ID by any of the infrastructure clients, it is sent to this blockchain. The blockchain consents to the records and accumulates them in a distributed ledger. This system possesses different types of clients in infrastructure analogous entry or exit tolls and agents of traffic [50].

In the unmanned aerial vehicle (UAV) application system, 6G technology can be used as incipient technology for communication in the UAV network. The challenges of UAVs is traffic in the air, path determination for the UAVs, their abundance in emergencies, which can be ameliorated by the distributed technology of the blockchain. This technology is the key for UAV communication in the 6G network. Each node in the network negotiates the record of data by furnishing adaptability, scalability, immutability, and transparency. It furnishes the decentralized ambient to the UAV, and 6G technology sponsors the reliability of vehicle-to-server and vehicle-to-everything connectivity. UAV roles the blockchain armature in 6G network can be customized in several ways, such as ameliorating network connectivity, emergency martialing in a catastrophe, security, and superintendence of pollution monitoring [16].

Blockchain accommodates UAV application in the 6G network, furnishing the accretion of data security in the air and stewardship of limited storage and processing power of UAVs in the acquisition, authentication, entry, transmission of data, as well as its automated verification in dissemination. Blockchain also enables hasty and secure communications and connections for UAV, downgrading the vulnerabilities of the system and physical hindrances in search and rescue, in securely furnishing medicines with no data manipulation, and in assisting with medical care. New support models and refinements in crop quality are feasible, and mid-air collision can be avoided with secured images from UAVs via aerial photography [16].

Smart logistics and transportation will enable the exploration of the Internet of Things (IoT) and blockchain technology. Transportation encompasses the conveyance of users and goods from one place to another. Logistics includes getting the product to the authorized carrier on time and in the correct quantity and good condition. The SmaTaxi is a smart transportation system that exploits IoT and blockchain with a decentralized ride-sharing network. This network registers private car owners as one node and processes the ledger. The smartphone with Internet connectivity helps in becoming a part of SmaTaxi [23].

This type of transportation requires the collection of data in real time, the analysis and transmission of data, and furnishing updates to the users. In smart logistics, the Saudi customs department integrates the IoT and blockchain. Appropriate parties can maneuver through the documents and data associated with a shipment in a single window, and the shipment may carry on to its destination on time. The system amends logistics by furnishing the respective information to the users and enhances distribution and delivery through unbroken consignments [23].

The sixth generation has ultrahigh data rates that enable more nodes to be connected to the network, and deploying the blockchain technology makes the vehicular ad hoc network more secure for transportation in the future.

7.6 CONCLUSION

Blockchain will revolutionize maneuvering and sectors in big data and universal connectivity in the sixth generation. Dynamic information access with ultrahigh data rates in 6G and handling variable devices like unmanned aerial vehicles with artificial intelligence can be deployed with blockchain technology. This deployment helps provide solutions for the security challenges faced due to the trust level of centralized authority for transactions in VANET. 6G sidesteps the knot of spectrum requirements in 5G and offers secure, low-cost, smart, and efficient spectrum utilization. Technologies like automation, robotics, and autonomous systems will be facilitated in the sixth generation by direct communication between intelligent vehicles and servers. 6G also facilitates the subsistence of unmanned aerial vehicles that can be hooked into wireless communication, operating with higher data rates than traditional base stations.

This chapter discussed the basic features of vehicular networks and the design goals of the communication model. It also depicted the system model with three main facets: OBU, RSU, and TA. Also, it illustrated the types of security attacks based on confidentiality between nodes, availability of resources, the authenticity of conveyance, the integrity of messages, and nonrepudiation of nodes. Finally, this chapter discussed how a blockchain is suitable for VANET in 6G by furnishing node trustworthiness in a distributed ledger for security attacks in a network.

NOTE

1 Doper accouterment refers to the additional user in the network.

REFERENCES

[1] Elmeadawy, S., Shubair, R. (2020) 6G Future Wireless Communications: Enabling Technologies, Opportunities, and Challenges:1–9. arXiv:2002.06068v1.
[2] Chatterjee, R., Chatterjee, R. (2017) An Overview of the Emerging Technology: Blockchain. Paper presented in 2017 3rd International Conference on

Computational Intelligence and Networks (CINE), Odisha, 126–127. doi: 10.1109/CINE.2017.33.

[3] Baygin, N., Baygin, M., Karakose, M. (2019) Blockchain Technology: Applications, Benefits and Challenges. Paper presented in 2019 1st International Informatics and Software Engineering Conference (UBMYK), Ankara, Turkey, 1–5. doi: 10.1109/UBMYK48245.2019.8965565.

[4] Holbl, M., Kompara, M., Kamisalic, A., Nemec Zlatolas, L. (2018) A Systematic Review of the Use of Blockchain in Healthcare. *Symmetry* 10:1–22. doi: 10.3390/sym10100470.

[5] Sabry, S.S., Kaittan, N.M., Ali, I.M. (2019) The Road to the Blockchain Technology: Concepts and Types. *Periodicals of Engineering and Natural Sciences* 7(4):1821–1832. doi: 10.21533/pen.v7i4.935.

[6] Awais Javed, M., Ben Hamida, E., Znaidi, W. (2016) Security in Intelligent Transportation Systems for Smart Cities: From Theory to Practice. *Sensors* 16(6):1–25.

[7] Li, Q., Wang, F., Wang, J., Li, W. (2019) LSTM-Based SQL Injection Detection Method for Intelligent Transportation System. *IEEE Transactions on Vehicular Technology* 68(5):4182–4191. doi: 10.1109/TVT.2019.2893675.

[8] Zeadally, S., Hunt, R., Chen, Y., et al. (2012) Vehicular Ad Hoc Networks (VANETs): Status, Results, and Challenges. *Telecommunication Systems* 50:217–241. doi: 10.1007/s11235-010-9400-5.

[9] Mathew, T.V. (2014) Intelligent Transportation System—II. *Transportation System Engineering* 18:1–49.

[10] Chen, Q., Sowan, A.K., Xu, S. (2018) A Safety and Security Architecture for Reducing Accidents in Intelligent Transportation Systems. Paper presented in 2018 IEEE/ACM International Conference on Computer-Aided Design (ICCAD), San Diego, CA, 1–7. doi: 10.1145/3240765.3243462.

[11] Kim, S. (2019) Impacts of Mobility on Performance of Blockchain in VANET. *IEEE Access* 7:68646–68655. doi: 10.1109/ACCESS.2019.2918411.

[12] Yang, W. (2013) Security in Vehicular Ad Hoc Networks (VANETs). *Wireless Network Security*. doi: 10.1007/978-3-642-36511-9_6.

[13] Mohanty, S., Jena, D. (2012) Secure Data Aggregation in Vehicular-Adhoc Networks: A Survey. *Procedia Technology* 6:922–929. doi: 10.1016/j.protcy.2012.10.112.

[14] Dai Nguyen, H.P., Zoltán, R. (2018) The Current Security Challenges of Vehicle Communication in the Future Transportation System. Paper presented in 2018 IEEE 16th International Symposium on Intelligent Systems and Informatics (SISY), Subotica, 000161–000166 September 2018. doi: 10.1109/SISY.2018.8524773.

[15] Lu, Z., Qu, G., Liu, Z. (2019) A Survey on Recent Advances in Vehicular Network Security, Trust, and Privacy. *IEEE Transactions on Intelligent Transportation Systems* 20(2):760–776. doi: 10.1109/TITS.2018.2818888.

[16] Aggarwal, S., Kumar, N., Tanwar, S. (2020) Blockchain Envisioned UAV Communication Using 6G Networks: Open issues, Use Cases, and Future Directions. *IEEE Internet of Things Journal*. doi: 10.1109/IoT.2020.3020819.

[17] Gupta, R., Tanwar, S., Kumar, N., Tyagi, S. (2020) Blockchain-Based Security Attack Resilience Schemes for Autonomous Vehicles in Industry 4.0: A Systematic Review. *Computers & Electrical Engineering* 86:1–15. doi: 10.1016/j.compeleceng.2020.106717.

[18] Zhang, X.D., Li, R., Cui, B. (2018) A Security Architecture of VANET Based on Blockchain and Mobile Edge Computing:258–259. doi: 10.1109/HOTICN.2018.8605952.

[19] Chen, Z., Zhou, K., Liao, Q. (2019) Quantum Identity Authentication Scheme of Vehicular Ad-Hoc Networks. *International Journal of Theoretical Physics* 58:40–57. doi: 10.1007/s10773-018-3908-y.

[20] Nkenyereye, L., Liu, C.H., Song, J.S. (2019) Towards Secure and Privacy Preserving Collision Avoidance System in 5G Fog Based Internet of Vehicles. *Future Generation Computer Systems* 95:488–499. doi: 10.1016/j.future.2018.12.031.

[21] Hewa, T., Gür, G., Kalla, A., Ylianttila, M., Bracken, A., Liyanage, M. (2020) The Role of Blockchain in 6G: Challenges, Opportunities and Research Directions 2nd 6G Wireless Summit (6G SUMMIT), Levi, Finland, 1–5. doi: 10.1109/6GSUMMIT49458.2020.9083784.

[22] Aljunaid, M.A.H., Syed Mohd Nazri Mohd Warip, A.A., et al. (2018) Classification of Security Attacks in VANET: A Review of Requirements and Perspectives. *MATEC Web of Conferences* 150(5):1–7.

[23] Humayun, M., Jhanjhi, N., Hamid, B., Ahmed, G. (2020) Emerging Smart Logistics and Transportation Using IoT and Blockchain. *IEEE Internet of Things Magazine* 3(2):58–62, doi: 10.1109/IoTM.0001.1900097.

[24] Nampally, V., Sharma, M.R. (2017) A Survey on Security Attacks for VANET. *International Journal of Computer Science and Mobile Applications* 5(10):58–70.

[25] Mishra, R., Singh, A., Kumar, R. (2016) VANET Security: Issues, Challenges and Solutions. Paper presented in 2016 International Conference on Electrical, Electronics, and Optimization Techniques (ICEEOT), 1050–1055. doi: 10.1109/ICEEOT.2016.7754846.

[26] Durga, R., Poovammal, E. (2020) Generation of RAESSES Hash Function for Medical Blockchain Formation Based on High Dynamic Chaotic Systems. *International Journal of Advanced Science and Technology* 29(6):8427–8440.

[27] Buinevich, M., Vladyko, A. (2019) Forecasting Issues of Wireless Communication Networks' Cyber Resilience for an Intelligent Transportation System: An Overview of Cyber Attacks. *Informations* 10(27):1–22. doi: 10.3390/info10010027.

[28] Thilak, K.D., Amuthan, A. (2016) DoS Attack on VANET Routing and Possible Defending Solutions—A Survey. Paper presented in 2016 International Conference on Information Communication and Embedded Systems (ICICES), 1–7. doi: 10.1109/ICICES.2016.7518892.

[29] Singh, K., Sharma, S. (2020) Advanced Security Attacks on Vehicular Ad Hoc Networks (VANET). *International Journal of Innovative Technology and Exploring Engineering (IJITEE)* 9(2):3057–3064. doi: 10.35940/ijitee.B7687.129219.

[30] Gautham, P.S., Shanmughasundaram, R. (2017) Detection and Isolation of Black Hole in VANET. Paper presented in 2017 International Conference on Intelligent Computing, Instrumentation and Control Technologies (ICICICT), Kannur, 1534–1539. doi: 10.1109/ICICICT1.2017.8342799.

[31] Baiad, R., Otrok, H., Muhaidat, S., Bentahar, J. (2014) Cooperative Cross Layer Detection for Blackhole Attack in VANET-OLSR. Paper presented in 2014 International Wireless Communications and Mobile Computing Conference (IWCMC), Nicosia, 863–868. doi: 10.1109/IWCMC.2014.6906469.

[32] Verma, S., Mallick, B., Verma, P. (2015) Impact of Gray Hole Attack in VANET. Paper presented in 1st International Conference on Next Generation Computing Technologies (NGCT), Dehradun, 127–130. doi: 10.1109/NGCT.2015.7375097.

[33] Kaur, G. (2016) A Preventive Approach to Mitigate the Effects of Gray Hole Attack Using Genetic Algorithm. Paper presented in 2016 International Conference on Advances in Computing, Communication, & Automation (ICACCA) (Spring), Dehradun, 1–8. doi: 10.1109/ICACCA.2016.7578899.

[34] Mejri, M., Ben-Othman, J. (2017) GDVAN: A New Greedy Behavior Attack Detection Algorithm for VANETs. *IEEE Transactions on Mobile Computing* 16(03):759–771. doi: 10.1109/TMC.2016.2577035.

[35] Lo, N., Tsai, H. (2007) Illusion Attack on VANET Applications—A Message Plausibility Problem. 2007 IEEE Globecom Workshops, Washington, DC, 1–8. doi: 10.1109/GLOCOMW.2007.4437823.

[36] Muhamad, A., Elhadef, M. (2019) Sybil Attacks in Intelligent Vehicular Ad Hoc Networks: A Review. In: Park, L., Loia, V., Choo, K.K., Yi, G. (eds) *Advanced Multimedia and Ubiquitous Engineering. MUE 2018, FutureTech 2018. Lecture Notes in Electrical Engineering*, Vol. 518. Springer, Singapore, 547–555. doi: 10.1007/978-981-13-1328-8_71.

[37] Ali, S., Nand, P., Tiwari, S. (2017) Secure Message Broadcasting in VANET Over Wormhole Attack by Using Cryptographic Technique. Paper presented in 2017 International Conference on Computing, Communication and Automation (ICCCA), Greater Noida, 520–523. doi: 10.1109/CCAA.2017.8229856.

[38] Bittl, S., Gonzalez, A.A., Myrtus, M., Beckmann, H., Sailer, S., Eissfeller, B. (2015) Emerging Attacks on VANET Security Based on GPS Time Spoofing. Paper presented in 2015 IEEE Conference on Communications and Network Security (CNS), Florence, 344–352. doi: 10.1109/CNS.2015.7346845.

[39] Grover, J., Gaur, M.S., Laxmi, V. (2011) Position Forging Attacks in Vehicular Ad Hoc Networks: Implementation, Impact and Detection. Paper presented in 7th International Wireless Communications and Mobile Computing Conference, Istanbul, 701–706. doi: 10.1109/IWCMC.2011.5982632.

[40] Sarencheh, A., Asaar, M.R., Salmasizadeh, M., Aref, M.R. (2017) An Efficient Cooperative Message Authentication Scheme in Vehicular Ad-hoc Networks. Paper presented in 2017 14th International ISC (Iranian Society of Cryptology) Conference on Information Security and Cryptology (ISCISC), Shiraz, 111–118. Doi: 10.1109/iscisc.2017.8488367.

[41] Chhoeun, S.A., Ayutaya, K.S.N., Charnsripinyo, C., Chamnongthai, K., Kumhom, P. (2009) A Novel Message Fabrication Detection for Beaconless Routing in VANETs. Paper presented in 2009 International Conference on Communication Software and Networks, Macau, 453–457. doi: 10.1109/ICCSN.2009.29.

[42] Zhang, C., Chen, K., Zeng, X., Xue, X. (2018) Misbehavior Detection Based on Support Vector Machine and Dempster-Shafer Theory of Evidence in VANETs. *IEEE Access* 6:59860–59870. doi: 10.1109/ACCESS.2018.2875678.

[43] Abbas, S., Faisal, M., Ur Rahman, H., Khan, M.Z., Merabti, M., Khan, A.U.R. (2018) Masquerading Attacks Detection in Mobile Ad Hoc Networks. *IEEE Access* 6:55013–55025. doi: 10.1109/ACCESS.2018.2872115.

[44] Rodrigues, B., Bocek, T., Lareida, A., Hausheer, D., et al. (2017) A Blockchain-Based Architecture for Collaborative DDoS Mitigation with Smart Contracts. In:

IFIP *International Conference on Autonomous Infrastructure, Management and Security.* Springer, Cham, 16–29.

[45] Yang, H., Zheng, H., et al. (2017) Blockchain Based Trusted Authentication in Cloud Radio Over Fiber Network for 5G. Paper presented at the 16th International Conference on Optical Communications and Networks, IEEE, 1–3.

[46] Shrestha, R., Bajracharya, R., Shrestha, A.P., Nam, S.Y. (2019) A New Type of Blockchain for Secure Message Exchange in VANET. *Digital Communications and Networks* 6(2):177–186. doi: 10.1016/j.dcan.2019.04.003.

[47] Leiding, B., Memarmoshrefi, P., Hogrefe, D. (2016) Self-Managed and Blockchain-Based Vehicular Ad-Hoc Networks. Proceedings of the 2016 ACM International Joint Conference on Pervasive and Ubiquitous Computing: Adjunct (UbiComp'16), Association for Computing Machinery, New York, 137–140. doi: 10.1145/2968219.2971409.

[48] Khan, A.S., Balan, K., Javed, Y., Tarmizi, S., Abdullah, J (2019) Secure Trust-Based Blockchain Architecture to prevent Attacks in VANET. *Sensors* 19:1–27. doi: 10.3390/s19224954.

[49] Rodrigues, B., Bocek, T., Lareida, A., Hausheer, D., Rafati, S., Stiller, B. (2017) A Blockchain-Based Architecture for Collaborative DDoS Mitigation with Smart Contracts. In: Tuncer, D., Koch, R., Badonne, R., Stiller, B. (eds) *Security of Networks and Services in an All-Connected World* 10356:16–29. doi: 10.1007/978-3-319-60774-0_2.

[50] Kulathunge, A.S., Dayarathna, H.R.O.E. (2019) Communication Framework for Vehicular Ad-Hoc Networks Using Blockchain: Case Study of Metro Manila Electric Shuttle Automation Project. Paper presented in 2019 International Research Conference on Smart Computing and Systems Engineering (SCSE), Colombo, Sri Lanka, 85–90.

Chapter 8

Blockchain and 6G Communication with Evolving Applications

Forecasts and Challenges

Preeti Yadav,[1] Mano Yadav,[2] Omkar Singh,[3]
and Vinay Rishiwal[4]

CONTENTS

[1,4] Department of CS & IT, MJP Rohilkhand University, Bareilly, India
[2] Department of Computer Science, Bareilly College, Bareilly, India
[3] National Institute of Fashion Technoloy, Patna, India
Corresponding author: omkar650@gmail.com

DOI: 10.1201/9781003264392-8

8.1 INTRODUCTION

Clearly, numerous facets of our atmosphere can be observed as informal ecospheres [1].

From broadcasting networks to the worldwide environment, from highway transportation networks to stock markets, from organic to communal systems, enormously interrelated and cooperating components create comparatively energetic systems in an ecosphere. These methods can be categorized as multifarious systems [2]. Multifarious systems investigation can be reflected as a science that educates us in what way the fundamentals of a system advance its cooperative performances and how the system interrelates with its surroundings. To comprehend the comportment of a given system, it is needed to originally comprehend the comportment of its fundamentals and how they affect the whole system's comportment. Multifarious methods and their anticipated comportments often include allusions to adaptability, emergence, self-organization, development, robustness, resilience, regionalization, and speed. Work emphasizes the fundamental features of multifarious methods in this perspective, which can be classified as reorganized [3]. As linkages of interrelating objects, multifarious methods are premeditated empirically through the support of a rapid upsurge of available statistics in distinct areas. Concomitantly, these distinct areas seem to pose numerous novel and important hypothetical queries. Movable communication systems, particularly in 5G and the upcoming 6G, are characteristic illustrations of systems that increase quickly [4]. Grasping their difficulty in terms of heterogeneous fundamentals and a large level of independence is a formidable obstacle, intimidating enough to materially interrupt the revolution. Conceiving, monitoring, demonstrating, and observing the performance of such networks are the major challenges that need to be tackled. We require novel models as we quickly transition from networks based on fastened classified or semiclassified edifices to exposed and dispersed informal networks [5]. From the broadcasting networks viewpoint, the crucial issue is to acquire how to project these networks that can self-establish, adapt, and improve their connections and tasks in an unremitting and healthy way to fulfill customer requests. The multifarious systems thus acquired can offer replicas, concepts, tools, and methods to permit an honorable project technique to be developed to tackle this main issue [6]. Movable broadcasting systems, particularly the 5G system and forthcoming 6G systems, are extra intricate and diverse. The emblematic process of these systems occurs through more solid placements, additional proper positions, uncountable users, and novel tools that are predicted to be announced in 6G systems such as machine learning (ML), artificial intelligence (AI), terahertz (THz) band broadcasting, and much more [7]. It is again impaired by a tendency to develop software for networking specifications and the energetic adaptation of networked amenities. The multifarious schemes concept could develop a valuable and operative contrivance proficient enough to be prototypical at a certain network performance [8].

This chapter presents multifarious systems from a broadcasting network viewpoint, enlightening issues, and challenges for 6G networks. The work focuses on 5G/6G communication systems, with the key focus on 6G by evolving 5G. The rest of the chapter presents 6G architecture, a technology used, and applications.

8.1.1 Motivation

The world has perceived novel network generations each decade since the 1980s. Every novel generation has provided better features than the previous one. The newly evolving generation, 5G, has numerous innovative features. 5G is expected to have numerous deficiencies when associated with further modern ICT substitutes. These deficiencies will be a key motivation for subsequent new network generation called the sixth generation of mobile communication (6G).

8.1.2 Research Contribution

In this chapter, the key research contributions are as follows:

- Upcoming challenges in 6G communications for canvassers are deliberated.
- Various 6G application areas are also presented in this survey.
- The key focus is associated with 6G architecture which includes protocols stacks, coverage, and artificial intelligence.
- Comprehensive analysis on development toward 6G communication is discussed.

- Diverse network generations with specifications and growths are elaborated.
- The main goal of this chapter is to give an informative direction for consequent 6G communication research.

8.2 ADVANCEMENT OF MOBILE COMMUNICATION NETWORK

There has been a phenomenal advancement in mobile communication networks since the emergence of an analog communications network in the 1980s. This advancement is not a one-step process but consists of several generations with different standards, capacities, and techniques. The new generation has been introduced nearly every ten years [9]. The progress of the mobile network is shown in Figure 8.1 [2].

8.2.1 1G-3G Networks

In the 1980s, the 1G communication network was presented, intended for voice amenities, with a 2.4 kbps data rate. It implemented analog signals to communicate data without presentable wireless customary, subject to

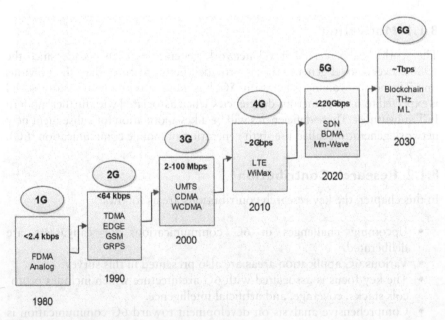

Figure 8.1 Mobile wireless networks advancement.

numerous shortcomings, including challenging handoff, squat broadcast competence, and no safety. In association with the 1G system, two generations were built on digital intonation tools, including TDMA and CDMA, with data rates up to 64 kbps, subsidiary voice, and SMS services. All network standards in the 2G epoch were GSM. 3G was projected in 2000 to have high data transferring speed and provided a 2 Mbps data transmission rate accessing the Internet [10].

8.2.2 4G Network

4G, introduced in the 2000s, is IP-based, which is proficient in providing data rates of high speeds up to 1 Gbit/s. It seemingly advanced phantom competence and decreased expectancy, cooperative necessities set by progressive applications such as DVD, TV content, and video conversation. Furthermore, 4G allowed great enough flexibility to offer wireless services anywhere and anytime, overcoming instinctive roaming, transversely topographical network restrictions. LTE-A and WiMAX are deliberated as 4G ethics. LTE assimilates prevailing and novel tools such as CoMP, MIMO, and OFDM [11].

8.2.3 5G Network

The 5G communication system accomplished preliminary rudimentary tests, h/w (hardware) amenities erection, and calibration procedure and was rapidly placed into commercial use. The objective of 5G was to create radical progress in connectivity, network reliability, latency, data, and energy proficiency. It customizes not just the novel range of microwave but also creatively customizes millimeter wave in the first period and enhances up to a 10 Gb data rate. 5G utilizes progressive admittance tools with BDMA FBMC. Numerous developing tools are united in 5G to achieve system concert: MIMO for capability upsurge, SDN for tractability in the system, D2D for phantom proficiency, ICN for decrease traffic in-network, and system slicing for rapid utilization of several services [12].

8.2.4 6G Network

As we know, 5G is ongoing the profitable distribution stage; research organizations everywhere in the world have paid courtesy to 6G network, which is deliberated to be installed sometime in the 2030s. Green sixth generation is anticipated to improve the concert of data broadcast up to 1 Tbps and to the ultralowest dormancy in microseconds. It topography includes THz broadcasting and latitudinal multiplexing, offering 1000 times the sophisticated capability of 5G systems. The solitary aim of 6G is to attain omnipresent connectivity by assimilating satellite broadcasting systems and underwater broadcasting in

Table 8.1 Use-Cases for 6G Networks

Use-case (proficiency)	5G	6G
Peak rate and capacity	Low resolution High-level task	High resolution Multisensory Detailed-level task Codesign
Telepresence	High video quality Limited scale	Mixed reality Holographic
Positioning and sensing	External sensing limited automation	Integrated radio sensing Fully automated
Time synchronization	Microsecond-level task	Higher precision Nanosecond-level tasks
Real-time multisensory mapping and rendering	No	Yes
Peak rate and capacity	No	Yes
Backscatter communications	No	Yes
Low-latency D2D	May be	Yes
Biosensors and AI	Limited	Yes

order to offer global exposure. Power reaping tools and customizing of innovative resources will significantly achieve the network power effectiveness and help in understanding supportable green systems [13].

Some novel characteristics that will also become significant for 6G's specified prospective use-cases are described [3] in Table 8.1.

8.3 RELATED WORK

Lv et al. [14] proposed a sensible elucidation to decrease signal intrusion for better transmission of associated signals. Furthermore, node info in sensors is handled by edge and fog computing; network broadcasting excellence is also arbitrated by transmission power utilization and packet failure rate. Dual-channel design can transfer control communications distinctly, minimize single-channel traffic load, and evade crashes amid control and sensor messages. It achieves the performance of transmission information of 6G/IoE. The utility representation of the proposed model is given as:

$$I_{utility} = \frac{I_{times}}{T_{current} - T_{first}} \quad (8.1)$$

Lv et al. [14] have summarized the notations and symbols used in Eq. (8.1).

The broadcast strategy has been professed as an arbitrarily performing object between receiver and transmitter by Basar et al. [15], which reduces the excellence of the acknowledged signal because of irrepressible connections of communicated radios by nearby substances. Instead, system machinists control the sprinkling, response, and bending features of radio waves by incapacitating the undesirable properties of the usual propagation in wireless. Baseband signal strength is used as follows:

$$r(t) = \frac{\lambda}{4\pi} \left(\frac{e^{\frac{-j2\pi l}{\lambda}}}{l} + \frac{R \times e^{\frac{-j2\pi(r_1+r_2)}{\lambda}}}{r_1 + r_2} \right) x(t) \tag{8.2}$$

Basar et al. [15] have described specific symbols and notations mentioned in Eq. (8.2).

Sim et al. [16] presented DNN construction, clarified how to evaluate PDP of sub-6 GHz station, which is implemented as input of DNN, and then authenticated its recital by actual surroundings—built 3D ray—outlining imitations over midair experimentations using millimeter wave archetype. The mth ray is used to design this prototype as:

$$h_m^{sub} = \sqrt{P_m} e^{j\phi_m^{sub}} F_{TX}^{sub}(\theta_m^{ZoD}, \phi_m^{AoD}) \tag{8.3}$$

Sim et al. [16] have summarized the symbolizations used in Eq. (8.3).

An IAP-SP reduces operational difficulty in preserving precise channel retrieval. Using the assessed station, the data rate expansion delinquent is expressed and is transformed in a separate-stage modification exploration problematic. Ma et al. [17] proposed the comprehensive exploration scheme to acquire an optimum broadcast rate and to tolerate a tremendously high operational load. Formerly, a local exploration scheme was developed to minimize the separate-stage IRS candidates through experiences of palpable performance failure. The transition contribution is implemented to develop the schemes.

$$\sigma = \frac{2e^2}{\pi \hbar^2} k_B T . \ln\left[2\cosh\left(\frac{E_F}{2k_B T} \right) \right] \frac{i}{\omega + i\tau^{-1}} \tag{8.4}$$

Ma et al. [17] have elaborated the notations and symbols used in Eq. (8.4).

Ahmad et al. [18] proposed the TPCSS technique for 6G solutions. A complexity exploration is conceded to evaluate the effect of medium and laser constraints on TPCS and comparative TPA drifts. Reconnoitered laser constraints are minimum power and beat size. Variable medium constraints are peroxide elucidation attentions and illustration length. A telemetry and command processing system (TPCS) is originated in order to control variation in minimum power. The best-fitted data can be calculated using TPA as follows:

$$T(z) = \sum_{n=0}^{\infty} \frac{(-q_0)^n}{(n+1)^{3/2}(1+x^2)^n} \tag{8.5}$$

Ahmad et al. [18] have described the symbolizations used in Eq. (8.5).

The SERS technique is proposed by Sykam et al. [19] for uncovering the squat attentiveness of peroxide molecules to perceived water prevention. Here is a low-cost, quick, and effective method for the manufacture of EG below microwave radioactivity in 1 min at 800 W and with outstanding adsorption substantial for R6G. The consequence of adsorption procedure constraints, including contact time, pH, isotherm replicas, and kinetic reproductions on rinse elimination under aqueous resolutions, was explored. Adsorption capability can be intended as:

$$q_e = (C_0 - C_e)\frac{V}{m} \tag{8.6}$$

Sykam et al. [19] has described the notations used in Eq. (8.6).

Chhabra et al. [20] stated the amalgamation of merged polyaniline using PbS QDs, which was successively engaged for snap catalysis of peroxide and Rh-6G. This PbS/PANI was amalgamated by commissioning biochemical oxidative polymerization in the occurrence of PbS QDs. The amalgamation was examined by X-ray precipitate deflection, FTIS, broadcast microscopy electron, and UV spectroscopy. NOMA and mmWave NOMA for a forthcoming B5G and 6G network explored by Zhu et al. [21]. The unique characteristic of mmWave NOMA is to receive/transmit beams using large, staged arrays. Yazhar et al. [22] projected 5G NR with possible waveform edifices to estimate waveform constraints in 6G. Various waveform constraint selections are forthcoming. TPs will implement these waveform constraint possibilities, even though they are transmitted to dissimilar users, using optimum resource distribution pronouncements. Goul et al. [23] measured CVD on tasters of dissimilar attentions of R6G on grapheme/AuNP substrates by implementing minimum energy. Limited component simulations were done for a network using hemi-ellipsoids in numerous circumstances, including an R6G investigative wrapper with superficial graphene nanoparticles. Graphene exists amid nanoparticles produced by redshift in plasmatic timber rate and superficial grapheme reduced electric arena.

8.4 TECHNOLOGIES USED IN 6G

8.4.1 Spectrum Communication Technique

Spectrum is the underpinning of broadcasting. Since the increase of communication in the 1980s, we have perceived marvelous development of spectrum possessions in each novel generation because of endless recreation for data rates—the foremost marks of 6G providing Tbps accumulated bit. Terahertz (THz) and perceptible light are two striking spectrums [24].

8.4.1.1 THz Communication

THz is a spectral ensemble amid warm and visual ensembles with a frequency from 0.1 THz to 10 THz. Excluding plentiful immature spectrum assets, THz inspires numerous sole features for forthcoming networks [25].

8.4.1.2 Perceptible Light Communication

OWC is reflected as a harmonizing tool for RF-based broadcasting. The frequency limit contains ultraviolet, perceptible light, and infrared spectrum. The perceptible nimble spectrum is the furthermost auspicious continuum of OWC because of scientific improvements and extensive implementation of LED. LED is differentiated from traditional lighting expertise due to rapid switching to dissimilar light strength echelons, allowing encoded data in produced light in diverse ways. Perceptible light broadcasting receives the full benefit of LED to attain the double goals of whirlwind fast and very large data transmission [26].

8.4.2 Fundamental Techniques

8.4.2.1 Blockchain for Decentralized Security

Blockchains are distributed record-based catalogs, and transactions can be steadily recorded and restructured independently of central mediators. The intrinsic geographies of blockchain, including dispersed interference with confrontation and secrecy, make it perfect for numerous applications. Blockchain is reflected as a subsequent revolution for forthcoming broadcasting tools. It assures tougher safety topographies during broadcast since it allows numerous system objects to strongly deter precarious conditions. Blockchain also delivers numerous assistances in resource instrumentation and system access [27].

8.4.2.2 Flexible and Intelligent Material

Notwithstanding marvelous achievements in broadcast systems in previous years, a concert of customary semiconductor resources, such as silicon, appears to meet its restrictions and resources with improved large-frequency and high-temperature features as an imperative necessity for ultrahigh broadcasting. Silicon and grapheme are used to project succeeding generation broadcasting devices [28].

8.4.2.3 Energy Harvesting and Management

Reliable calculation strains for AI dispensation and enhancing the explosion of IoT strategies are posing important challenges to the power efficacy of broadcast equipment. Consequently, power-competent broadcast tools will shine in 6G having a shorter broadcast distance. Many exertions have been expended on power reaping and organization investigations in previous years. The SR tool provides a conceivable elucidation of power delinquence, which assimilates unreceptive backscatter strategies with a vigorous broadcast network [29].

8.4.3 New Communication Paradigm

8.4.3.1 Molecular Broadcast

A novel broadcast model, originating in the environment, is a conceivable elucidation that implements biological signals to transmit data: molecular broadcast (MB). In MB, biological signals are characteristically minor elements of nanometers to micrometers that, to some extent, include phospholipid vesicles and atoms, which typically broadcast in gases. In terms of radio broadcasting, MB has definite benefits in the micro- and macrogauges [30].

8.4.3.2 Quantum Broadcast

Quantum broadcast (QB) is an auspicious broadcast prototype of unrestricted safety. The ultimate variance amid quantum broadcast and traditional binary-based broadcast can be sensed on-site. Data is encrypted in the quantum stage with photons/quantum atoms and cannot be retrieved or replicated. Attackers are denied interfering due to quantum ideologies, including the association of tangled atoms and unchallengeable law. Moreover, QB can achieve its data rates because of the superposed environment of cubits. Another striking fact of QB is its enormous prospective in long-distance broadcast. QR is a serious strategy for a long-distance universal quantum system. It is proficient in separating the QB into smaller intermediary sections and modifying photon damage and process faults [31].

8.5 CHALLENGES IN 6G COMMUNICATION

Through the previous two periods, the cellular systems tools developed from 1G through 5G. 5G networks were anticipated to in use up to 2020, and communal investigation has started into how sophisticated subsequent generation of systems will be. A large number of research articles already available are introducing 6G systems with features. The forthcoming 6G network will face various key challenges, and these are discussed here [32].

8.5.1 Dynamic Topology

The network topology in the sixth generation is predictably dynamic. Because every operator of system/smart devices forming IoT systems will be associated actively with a network and will need to offer preeminent QoS in an instant, networks will need to be extremely dynamic. Drones, UAVs, drones, radar, and satellite broadcast will be quick-acting nodes that should support complexity. Essential and appropriate prototypes will be needed so that devices can be converted rapidly to subnetworks while maintaining network reliability [33].

8.5.2 THz Frequencies

The need for sophisticated data rates and large spectral power proficiency call for frequencies outside the millimeter wave at terahertz (THz) capacity. Thz provides tiny cell enlargement whose range is limited up to meters. Minute cells will initiate much thicker positioning, whose dynamism investigators need to develop novel traffic flow supervision methods, movement controlling, mobbing mechanism algorithms, and much more [34].

8.5.3 Access Network for Backhaul Traffic

Network tools will necessitate an enormous escalation in data evolution and reduction in network access for backhaul to be effective. Research enhancement in advanced bands such as D-Band, including 60 GHz ranges, is accessible. FSO and quantum broadcast also reflect 6G backhaul to come up with the necessities. Drones and telluric stations may involve satellite connections with squat trajectory satellites and CubeSats for providing backhaul sustenance and attention to upsurge in large area [35].

8.5.4 Artificial Intelligence and Machine Learning

Because of the difficulty of 6G systems, AI will be predictably included for the efficacious and effectual processes of such systems. AI has previously been implemented in broadcasting in each OSI layer. Concerning 6G systems, AI is a candidate to simplify the process and probably to lessen the difficulty. AI is predicted to play an important role in facets from semantic broadcasts, ML, and NN to the complete supervision of broadcasts [36].

8.5.5 Network Functions Virtualization and SDN

SDN and NFV are the tools that depend on virtualization. The intent of these tools is to permit system project and structure in s/w (software) execution by fundamental s/w crosswise generic h/w platforms and strategies. SDN emphasizes the disentangling of system mechanism tools from network advancing tools, whereas NFV eliminates system advancing and former informal tools from the h/w, prior to softwarization of system tools. In system amenities adaptation, the execution of functioning processes involved in the manipulation, making, and distribution of E2E services, NFV and SDN will improve adaptively in 6G networks [37].

8.5.6 Blockchain

A blockchain is a tool that is intended to sustain 6G systems. Subsequently, it is a tool that can expressively underwrite an enormous amount of statistics for an organization in 6G broadcast networks. Blockchain is attained by peer-to-peer systems and operates dependently of an integrated authority. Blockchain is likely to deliver numerous automatic amenities, including the interoperability of transverse devices, immense traceability of data, and communications of different IoT schemes [38].

8.5.7 Moving Networks

Users will require large, excellent, enormously enhanced Internet facilities on moving trains/vehicles/planes. Users request similar services as stationary substructure operators, and 6G systems should be capable to offer it. To tackle these demands, the idea of dynamic networks has been presented. Dynamic networks are a singular group of ad hoc networks in which, with their extremely unstable environment, quality disputes can be resolved despite the rapidity of vehicles [39].

8.5.8 Intelligent Surfaces

Intelligent surfaces encourage the development of h/w tools to achieve the range and power efficacy of WSN. MIMOs implement antennas composed of a large number of arrays to amend their radio transmission designs over time and prepare them for both broadcast and receiving. HSFs electromagnetically surf their broadcast surroundings for specified purposes through programmatically organized meta-exteriors. The RIS-associated idea encompasses RIS array components. Meta-exteriors are thin planar, artificial edifices that have recently permitted the comprehension of original electromagnetic and visual mechanisms by engineerable amenities [40].

8.6 6G APPLICATION AREAS

To enhance Internet implementation in rustic regions, users' require assistance in enhancing implementation and ultimately creating a feasible commercial situation, whose significance for rustic operators can be promoted. Moreover, operators need to adopt obtainable facilities in order to strengthen their petition and inspire users to increase the connectivity delivered to them. The various 6G applications areas are discussed here [41].

8.6.1 Health

In rustic zones, remote patients are entitled to excellent healthcare, and teleprescription is an appropriate response to tackle this issue. The main fitness hubs can be employed with regular stopovers; however, transfer to subordinate and tertiary fitness hubs does not include the facilities to precisely hand over patient information. Dissimilar methods are projected, wherein community employees assist patients in remote zones and interconnect with medics implementing hypermedia equipment relying on strengthened broadband reinforced by broadcasting [42].

8.6.2 Education

Education is a significant facility to deliver distant rustic zones. Distant teaching is delivered to rustic zones via virtual programs based on multimedia. The flipped-based classroom archetype was implemented for supporting online teaching. Candidates in remote areas are hampered by imperfect connectivity that delays the presentation of the classroom setup for video conferencing. A communicating teaching method implementing satellite TV networks is suitable for rustic regions. Because of the absence of dependable Internet broadcasting, learners have to use mostly customized tablets for interpretation of e-books [43].

8.6.3 Farming

Agriculture is a distinctive application for remote areas that can be assisted by Internet broadcasting or LAN. For instance, IoT-based devices are used for achieving exact irrigation in fields with imperfect access to irrigation sources. Trickle irrigation delivers a precise volume of water and improves sprayed fields near herbal areas [61]. Sensor nodes in IoT are projected for agriculture applications in rustic fields that are deprived of broadcasting or cloud communication. UAVs are implemented to gather capacities from sensors connected with the IoT and to communicate them to adjacent 5G base stations [44].

8.6.4 Financial Services

ATM development and use of POS tools pose many issues in rustic regions. A key challenge is the absence of dependable broadcasting for performing transactions through cards by users in real time. Improved safety can be attained with a smartphone, accompanied by a smart card, making use of a broadcasting approach. Though legitimate disputes must be dealt with, the utmost possibility is that of overdrawing on a bank account. Reliable ATMs might be enabled to collect the amount on the card and modify the balance during the withdrawal [45].

8.6.5 E-Commerce

A system for associating support assemblies in rustic regions has been developed. The main goal is to enable microbusinesspersons in remote areas to increase production and maintain frugality. The system implements IVR and mVAS to permit transactions among rustic shareholders to sustain microbusinesspersons in rustic regions and permit them to increase their commercial activities. A mobile-based system assumes that rustic inhabitants are adequately informed, that statements of revenues move to individuals' workstations, and that funds can be accessed via web services [46].

8.6.6 E-Government

Before instituting e-supervision in rustic regions, a bottom-up method is projected, wherein amenities connected to agriculture, health, and education should be placed on a standardized level prior to users joining the e-supervision plan. Otherwise, an e-supervision plan will not be successful unless the people in a rustic region are prepared for acceptance of the envisioned assistance [47].

8.6.7 Other Services

Kiosks offer a number of prospects for rustic regions. Job seekers can upload their information at a kiosk, and employers can upload openings. Similarly, kiosks can be implemented for selling/buying, where vendors can promote their products and services. All the kiosk are connected to servers through the Internet. Bus or train ticketing service is also a consideration in different areas; ticket retailing stations in remote areas provide unhurried connectivity to relevant servers, where the ML approach is implemented for handling such services [48].

The AI impact on 6G networks functions [49] is described in Table 8.2.

Table 8.2 AI Impacts on 6G Network Functions

Network layer	Functions	AI algorithms	Descriptions
Physical layer	Reliable data transmission Channel coding Modulation MIMO precoding OFDM channel estimation	DNNs Auto-encoder CNNs CCNNs K-means	Revamped end-to-end PHY architecture reduces complexity level of MIMO OFDM receiver. During high mobility, enhances PHY performance.
Data link layer	Frame following operations Flow control synchronization Data packet querying Scheduling Power control Error corrections Flow control	DNNs Supervised learning Deep learning Reinforcement learning Transfer learning Q-learning	Optimal user scheduling Improve network performance Channel estimation Traffic prediction Enhanced radio resources efficiency Optimal retransmission redundancy Reduction in retransmission overhead
Network layer	Responsible for routing Mobility RRC, correction, and load BS association BS clustering	DNNs K-means Unsupervised learning Supervised learning Reinforcement learning Q-learning	Optimizes serving cells Optimized multiple connectivities Mobility prediction Optimizes handover Optimized data transmission paths Optimizes BS clustering Controlling the size of the cluster in dynamic network scenarios

8.7 6G CASE STUDY: PROPOSED ARCHITECTURE

6G networks are anticipated to attain power proficient and communally unified wireless links worldwide. However, prevailing network design is incapable of assuring forthcoming application provision restraints, such as high throughput, low latency, and consistency. Consequently, forward investigation of forthcoming network outlines is essential. In this chapter, the architecture of the 6G network is not proposed to be formulated until 2030 [2]. It is impractical to precisely demonstrate what forthcoming 6G network architecture will be like, but we have presented 6G-associated architecture here. This architecture is of three basic categories: enhanced stratification, intelligent connection, and universal coverage. The stratification network view includes content-driven routing, management plane, dynamic spectrum access, fluid antenna, blockchain, new spectrum THz, and VLC. Intelligent connection is a

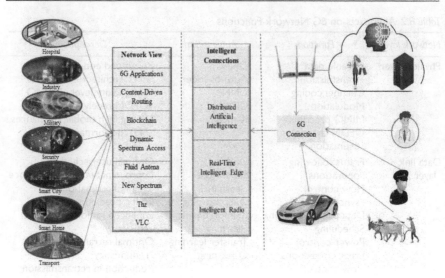

Figure 8.2 Network architecture for 6G.

significant view of this architecture introducing novel technologies in distributed artificial intelligence, intelligent radio, and real-time intelligent edge [50].
6G network coverage will include space network, aerial network, terrestrial
network, undersea network, and much more. HTS methods are proficient with
broadband network service similar to telluric facilities including evaluating
and bandwidth. An NGSO system provides minimum-latency, large-bit-rate
worldwide network access, as well as numerous satellite collections to encourage commercialization. LEO networks have the definitude in concept and
in simulating surroundings with laser and RF mechanisms to deliver lower-
latency broadcasting than terrestrial systems [51]. See Figure 8.2.

8.8 CONCLUSION

There has been nearly an exponential upsurge in broadcast data, particularly
multimedia statistics, and a rapid propagation of all types of smart devices
for subsequent broadcast development in anticipation of 6G. 6G communications are an auspicious, important upsurge in QoS and supportable future. In
this chapter, we have presented a detailed survey of 6G communications. We
described the development of network generations from 1G to 5G, as well
as the expansion drift of 6G at a certain level. Detailed studies of different
technologies used in 6G communication have been elaborated. This chapter
also delineated the upcoming challenges in 6G communication for investigators. Various 6G application areas were also presented in this survey. The key

focus was on 6G architecture, which includes protocols stacks, coverage, and artificial intelligence. The goal of this chapter is to give informative direction for consequent 6G communication research.

REFERENCES

1. Nawad, S.J, Sharma, S.K., Wyne, S. et al.: Quantum Machine Learning for 6G Communication Networks: State-of-the-Art and Vision for the Future. *DCN.* 7, 46317–46350 (2019).
2. Huang, T., Yang, W. and Wu, J.: A Survey on Green 6G Network: Architecture and Technologies. *IEEE Access.* 7, 175758–175768 (2019).
3. Vishwanathan, H. and Mogensen, P.E.: Communications in the 6G Era. *GCA.* 141, 57063–57074 (2020).
4. Sergiou, C., Lestas, M., Antoniou, P. et al.: Complex Systems: A Communication Networks Perspective Towards 6G. *Complex Networks Analysis and Engineering in 5G and Beyond Towards 6G.* 8, 89007–89030 (2020).
5. Yaacoub, E. and Aloumini, M.S.A.: A Key 6G Challenge and Opportunity—Connecting the Base of the Pyramid: A Survey on Rural Connectivity. *CNA.* 108, 533–582 (2020).
6. Wang, J., Qiu, C., Mu, X. et al.: Ultrasensitive SERS Detection of Rhodamine 6G and p-Nitrophenol Based on Electrochemically Roughened Nano-Au Film. *Talanta.* 19, 1–40 (2019).
7. Rappaport, T.S., Kanhere, O., Mandal, S. et al.: Wireless Communications and Applications Above 100 GHz: Opportunities and Challenges for 6G and Beyond. *Millimeter-Wave and Terahertz Propagation, Channel Modeling and Applications.* 141, 78729–78757 (2019).
8. Haas, H., Yin, L., Chen, C. et al.: Introduction to Indoor Networking Concepts and Challenges. *LiFi. Journal of Optical Communications and Networking.* 12, 190–203 (2020).
9. Lin, L. and Meng, W.: Convolutional-Neural-Network-Based Detection Algorithm for Uplink Multiuser Massive MIMO Systems. *EUSPN.* 8, 64250–64265 (2020).
10. Balasubramanian, K. and Swaminathan, H.: Highly Sensitive Sensing of Glutathione Based on Förster Resonance Energy Transfer Between MoS2 Donors and Rhodamine 6G Acceptors and Its Insight. *Sensors and Acuators.* 17, 1–41 (2017).
11. Barzan, M. and Hajiesmaeilbaigi, F.: Investigation the Concentration Effect on the Absorption and Fluorescence Properties of Rhodamine 6G Dye. *Optik.* 159, 157–161 (2019).
12. Dai, Y., Fei, Q., Shan, H. Y., et al.: Determination of Er3+ Using a Highly Selective and Easy-to-Synthesize Fluorescent Probe Based on Rhodamine 6G. *Arabian Journal of Chemistry.* 19, 1–27 (2019).
13. Manuel, M.L.R., Cantu, S., Lopez, E.P., et al.: Evaluation of Calcium Oxide in Rhodamine 6G Photodegradation. *Catalysis Today.* 17, 1–17 (2017).
14. Lv, J. and Kumar, N.: Software Defined Solutions for Sensors in 6G/IoE. *Computer Communications.* 20, 1–11 (2020).

15. Basar, E., Rosny, J.D., Aloumini, M.S., et al.: Wireless Communications Through Reconfigurable Intelligent Surfaces. *Computer Communications*. 7, 116753–116773 (2019).

16. Sim, M.S., Park, S.H. and Chae, C.B.: Deep Learning-Based mmWave Beam Selection for 5G NR/6G with Sub-6 GHz Channel Information: Algorithms and Prototype Validation. *Artificial Intelligence for Physical-Layer Wireless Communications*. 8, 51634–51646 (2020).

17. Ma, X., Chen, Z., Chi, Y., et al.: Joint Channel Estimation and Data Rate Maximization for Intelligent Reflecting Surface Assisted Terahertz MIMO Communication Systems. *Computer Networks*. 8, 99565–99581 (2020).

18. Ahmad, R., Rafique, M.S., Ajami, A. et al.: Influence of Laser and Material Parameters on Two Photon Absorption in Rhodamine B and Rhodamine 6G Solutions in MeOH. *Optik—International Journal for Light and Electron*. 183, 835–841 (2019).

19. Sykam, N., Jayram, N.D. and Rao, G.M.: Exfoliation of Graphite as Flexible SERS Substrate with High Dye Adsorption Capacity for Rhodamine 6G. *Applied Surface Science*. 471, 375–386 (2019).

20. Chhabra, V.A., Kaur, R., Walia, M.S., et al.: PANI/PbS QD Nanocomposite Structure for Visible Light Driven Photocatalytic Degradation of Rhodamine 6G. *Environmental Research*. 20, 1–32 (2020).

21. Zhu, L., Xia, X. and Wu, D.O.: Millimeter-Wave Communications with Non-Orthogonal Multiple Access for B5G/6G. *Millimeter-Wave Communications: New Research Trends and Challenges*. 7, 1161223–116132 (2019).

22. Yazar, A. and Arslan, H.: A Waveform Parameter Assignment Framework for 6G With the Role of Machine Learning. *VTS*. 1, 156–172 (2020).

23. Goul, R., Das, S., Liu, Q. et al.: Quantitative Analysis of Surface Enhanced Raman Spectroscopy of Rhodamine 6G Using a Composite Graphene and Plasmonic Au Nanoparticle Substrate. *Carbon*. 111, 386–392 (2017).

24. Ardakni, A.G. and Rafiepour, P.: Random Lasing Emission from WO3 Particles Dispersed in Rhodamine 6G Solution. *Physica B: Physics of Condensed Matter*. 18, 1–16 (2018).

25. Haroon, M., Wang, L., Yu, H., et al.: Synthesis of Carboxymethyl Starch-g-Polyvinylpyrolidones and Their Properties for the Adsorption of Rhodamine 6G and Ammonia. *Carbohydrate Polymers*. 186, 50–158 (2018).

26. Kang, H., Fan, C., Xu, H., et al.: A Highly Selective Fluorescence Switch for Cu2þ and Fe3þ Based on a New Diarylethene with a Triazole-Linked Rhodamine 6G Unit. *Tetrahedron*. 74, 4390–4399 (2018).

27. Malfatti, L., Suzuki, K., Erker, A., et al.: Photoluminescence of Zinc Oxide Mesostructured Films Doped with Rhodamine 6G. *Journal of Photochemistry and Photobiology A: Chemistry*. 18, 1–14 (2018).

28. Tang, F., Kawamoto, Y. and Kato, N., et al.: Future Intelligent and Secure Vehicular Network Toward 6G: Machine-Learning Approaches. *CAN*. 19, 1–16 (2019).

29. Viboonratansari, D., Pabchanda, S. and Prompinit, P.: Rapid and Simple Preparation of Rhodamine 6G Loaded HY Zeolite for Highly Selective Nitrite Detection. *Applied Surface Science*. 18, 1–33 (2018).

30. Wang, Z., Zhang, Q., Liu, J. et al.: A Twist Six-Membered Rhodamine-Based Fluorescent Probe for Hypochlorite Detection in Water and Lysosomes of Living Cells. *Analytica Chimica Acta*. **1082**, 116–125 (2019).

31. Wu, W.N., Wu, H., Zhong, R.B., et al.: Radiometric Fluorescent Probe Based on Pyrrole-Modified Rhodamine 6G Hydrazone for the Imaging of Cu2+ in Lysosomes. *Spectrochimica Acta Part A: Molecular and Biomolecular Spectroscopy*. **212**, 121–127 (2019).

32. Lu, Y. and Zheng, X.: 6G: A Survey on Technologies, Scenarios, Challenges, and the Related Issues. *Journal of Industrial Information Integration*. **19**, 1–52 (2020).

33. Wei, Y., Peng, M. and Liu, Y.: Intent-Based Networks for 6G: Insights and Challenges. *Digital Communications and Networks*. **5**, 1–11 (2020).

34. Sen, P., Pados, D.A., Batalama, S.N., et al.: The TeraNova Platform: An Integrated Testbed for Ultra-Broadband Wireless Communications at True Terahertz Frequencies. *Computer Networks*. **179**, 1–11 (2020).

35. Fu, Y., Doan, K.N. and Quek, T.Q.S.: On Recommendation-Aware Content Caching for 6G: An Artificial Intelligence and Optimization Empowered Paradigm. *Digital Communication and Networks*. **10**, 1–11 (2020).

36. Zhao, Y., Zhao, J., Zhai, W., et al.: A Survey of 6G Wireless Communications: Emerging Technologies. *Computer Networks*. **20**, 1–10 (2020).

37. Zhou, Y., Liu, L., Wang, L., et al.: Service Aware 6G: An Intelligent and Open Network Based on Convergence of Communication, Computing and Caching. *Digital Computer Networks*. **11**, 1–11 (2020).

38. Xu, H., Klaine, P.V., Onireti, O., et al.: Blockchain-Enabled Resource Management and Sharing for 6G Communications. *DCN*. **15**, 1–13 (2020).

39. Long, O., Chen, Y., Zhang, H., et al.: Software Defined 5G and 6G Networks: A Survey. *Mobile Networks and Applications*. **5**, 1–21 (2019).

40. Akyyildiz, J.F. and Nie, S.: 6G and Beyond: The Future of Wireless Communications Systems. *WCS*. **141**, 1–36 (2020).

41. Huang, T., Yang, W., Wu, J., et al.: A Survey on Green 6G Network: Architecture and Technologies. *GIT*. **7**, 175758–175768 (2019).

42. Shah, S.H.A., Balkrishan, S., Xin, L., et al.: Beamformed Mmwave System Propagation at 60GHz in an Office Environment. *ICC*. **11**, 1–7 (2020).

43. Rahman, T.A., Aziz, O.A., Hinda, M.N., et al.: Channel Characterization and Path Loss Modeling in Indoor Environment at 4.5, 28, and 38 GHz for 5G Cellular Networks. *IJAP*. **2018**, 1–14 (2018).

44. Mitra, R.N. and Agarwal, D.P.: G Mobile Technology: A Survey. *ICT Express*. **1**, 132–137 (2015).

45. Alsharif, M.H., Kelechi, A.H., Albreem, M.A., et al.: Sixth Generation (6G) Wireless Networks: Vision, Research Activities, Challenges and Potential Solutions. *MDPI*. **12**, 1–21 (2020).

46. Elmeadawy, S. and Shubair, R.M.: 6G Wireless Communications: Future Technologies and Research Challenges. *ICECTA*. 1–5 (2019).

47. Saad, W., Bennis, M. and Chen, M.: A Vision of 6G Wireless Systems: Applications, Trends, Technologies, and Open Research Problems. *Computer Science*. **21**, 1–26 (2019).

48. Zhang, Z., Xiao, Y., Ma, Z., et al.: 6G Wireless Networks: Vision, Requirements, Architecture, and Key Technologies. *EUSPN*. **14**, 1–14 (2019).
49. Siddiqui, M.H., Khurshid, K., Rashid, I., et al.: Artificial Intelligence based 6G Intelligent IoT: Unfolding an Analytical Concept for Future Hybrid Communication Systems. *WCSE*. 122–129 (2020).
50. Huang, T., Yang, Wu., Wu, J., et al.: A Survey on Green 6G Network: Architecture and Technologies. *Green Internet of Things*. **7**, 1–14 (2019).
51. Yang, P., Xiao, Y., Xiao, M., et al.: 6G Wireless Communications: Vision and Potential Techniques. *Licensed and Unlicensed Spectrum for Future 5g/B5g Wireless Systems*. **33**, 70–75 (2019).

Chapter 9

Emergence of Blockchain Applications with the 6G-Enabled IoT-Based Smart City

Meenu Gupta,[1] Chetanya Ved,[2] and Meet Kumari[3]

CONTENTS

[1] Department of Computer Science and Engineering, Chandigarh University, Punjab, India
[2] Department of Information Technology, Bharati Vidyapeeth's College of Engineering, Delhi, India
[3] Department of Electronics and Communication Engineering, Chandigarh University, Punjab, India
Corresponding author: meenu.e9406@cumail.in

DOI: 10.1201/9781003264392-9

9.1 INTRODUCTION

Cities are getting smarter and better planned as technology advances at a quickening pace. Due to the Internet, the living standards in cities are becoming more accessible and comfortable for people. A city can be described or designated as a smart city when information and communication technologies (ICT) is associated with its utility services and urban industrial sectors to eradicate resource waste, consumption, and overall cost. Smart cities are known for making comebacks in intelligent ways to mitigate different kinds of needs, such as livelihood, environmental safety, public and city services, industrial and commercial activities [1]. The motive behind developing such cities is to develop sustainable livelihoods for the citizens. It is observed that the living standards of those who live in such cities have improved. Giving them adequate training can uplift the industrial and economic sectors, thereby making them sustainable. The adoption of modern technologies and the development of a new standard operating procedure (SOP) for every sector will have the potential to revolutionize services across the board. The major problems faced by existing cities are urban population, illegal construction in open areas, shortage of electricity and gas supplies, insufficient medical services, and improper governance. Other problems faced by citizens are unemployment, transportation, the waste of nonrenewable natural resources, exploitation of the environment, lack of security, and surveillance. Since people face many problems living in rural areas due to which they are migrating to urban cities. It has been estimated that 30 million people will move from rural areas to urban cities by 2050 [2, 3]. To eliminate such problems, implementing several technologies to make processes efficient and effective is required. Since smart cities are data driven, it is important to transmit and exchange data with security. Processing and management of that data require high computational capabilities, which are further explained in this chapter.

This chapter mostly focuses on the amalgamation of several technologies working together to develop a sustainable environment. Section 2 is an elaborative literature survey. Section 3 describes the application of a smart city with an IoT-based framework used for data management and processing. Section 4 explains the implementation of blockchain with IoT in a smart city. Implementation of artificial intelligence in integration with an IoT-based network is explained in Section 5. Different generations of networks are described in Section 6. Case studies based on the UK and India's smart city initiatives are discussed in Section 7. The challenges and opportunities faced by modern technologies are described in Section 8.

9.2 LITERATURE SURVEY

Smart cities incorporate all these upcoming technologies, developed with a motive for sustainable development and raising the standard of living. It can be observed that every technology has its challenges and drawbacks to mitigate those obstacles. It is required to combine them in a network to overcome those hurdles. The latest literature survey of smart cities and their various frameworks is presented here.

Smart city architecture is dependent on the two major sectors: interconnectivity communication and technology (ICT) and telecommunication. A smart city's (SC's) business model is discussed, which outlines the benefits of SCs from an economic perspective with challenges [3]. An idea of SC, stating why people are more interested in living in a smart city, is discussed, and questions are answered in shifting this paradigm of people's mindset. Various applications such as smart economy, smart mobility, smart governance, smart living, etc. are revealed in this chapter [4]. The authors present a brief study about its architecture, applications, technologies, and challenges, which gives an overall glimpse of the present and upcoming scenarios of the SC [5]. Also, the author proposes a new type of SC architecture that can be viewed from the perspective of data; further categorization of functionality in SC has been done. Architecture is also revealing some design challenges and giving some insights into such cities [6]. In Arasteh et al. [7], a survey was done of those SCs where most applications are IoT based. The advancement of applications has been further discussed; a brief explanation is given about the implementation of IoT in each sector related to the smart city. The model of secured IoT architecture for smart cities is proposed. The secured architecture is based on black network, a trusted SDN controller, and unified registry and key management system. The integration of all these blocks will save the networks from the deception of cyberattacks [8]. A detailed study about implementing the machine learning model for blockchain-based smart applications to provide immunity against attacks is presented [8].

Further ML-based techniques discussed in this chapter such as support vector machine (SVM), clustering, Dl-based algorithms used for analyzing the blockchain-based network, have been presented. Furthermore, in this chapter, integrating both the technologies in applications of SCs is discussed [9]. A blockchain-based system is introduced to develop a trustless network for storing closed-circuit television camera (CCTV) recordings, allowing only respective authorities to verify that the videos recorded have been altered or not. Further, an immutable ledger will be generated of recorded video metadata that only authorized agencies can access [10]. Some areas of SCs where applications are based on artificial intelligence (AI) have been reviewed. Also, the author reviewed examples of such applications as traffic monitoring and prediction, analysis of social-related big data, routing, health quality management [11]. Again, a detailed survey about 5G technologies is mentioned in Gupta and Jha [12].

This focus is kept on 5G-based architecture; MIMO technology and device-to-device (D2D) are discussed in depth. Some challenges in implementing this next generation are given briefly, such as management of interference, sharing of spectrum among the cognitive radio, full-duplex systems. Working and practical implementation of cloud technology incorporation with 5G network are also described.

9.3 SMART CITY AND ITS DATA MANAGEMENT FRAMEWORK INTEGRATION WITH IoT-ENABLED NETWORK

The data management framework is required to manage and store a massive amount of produced data. Smart cities have multiple points of data generation sources, and utilizing it judiciously requires huge computing capabilities to process the data. This section of the chapter describes the steps of data processing and its application in various areas of the smart city [12].

These cities are designed keeping in mind certain factors such as population growth, environmental crisis, and maximum utilization of resources in upcoming years. Modern solutions and technologies are under development for these types of cities related to these fields, such as blockchain, Artificial Intelligence, financial technology (Fintech), etc. As a result, the city gets smarter and connected, which makes the environment more collaborative and efficient in terms of working and productivity.

Data is determined as having quantitative or qualitative values that can be transformed into more meaningful information through computation. It gives scientists, engineers, and inventors assistance in developing new ways to solve problems. A smart city is considered a hub of data which generates millions of data bits within nanoseconds from various types of sources. Maximum utilization of important capital (i.e., human capital) requires a medium to analyze or monitor data of each of the areas of the city so that related problems can be identified and rectified in a timely fashion. Extracting data from the environment and monitoring that data give us a pathway to solving such problems as pollution sensing and detection in cities, making the environmentalist engineers work on air filtering. It prevents people from becoming patients of chronic diseases such as asthma, pneumonia, and many other lung-related problems [12]. The related steps are as follows:

- Extraction/collection of data through a different type of sensors
- Data is classification, i.e., storing or visualizing it
- Cleaning of data and converting it into useful logical information
- Developing data sets from extracted data

In the modern era of the twenty-first century, the areas of existing cities are operational without concern for proper analyses of SOPs (standard operating procedures). As a result, the population faces major difficulties and problems. Broadly, there are seven areas of cities in which smart city have come up with innovation and technologies to adopt modern methods or ways to make procedures and operations smooth and efficient with the help of machines [13, 20]:

1. Security and surveillance
2. Logistics operations and management
3. Governance
4. Energy generation resource management (EGRM)
5. Transportation
6. Economy
7. Education

In Figure 9.1, it can be observed that no areas are connected. Every field performs its own activities. They do not collaborate, and the citizen cannot access information on any area of the existing city through one common platform. And the city generates redundant data that cannot be analyzed, and no solution can develop that possesses high enough accuracy to solve that particular problem.

Integration of different areas through a smart device–based city infrastructure will lead to developing a smart city, as shown in Figure 9.2. This type of infrastructure enables people to monitor, analyze, and access any of data via smart devices (i.e., smartphones, laptops, wearable devices). And this type of city generates data that can be easily convertible to information. It helps engineers, scientists, researchers to develop a solution collectively. Making cities smarter will ultimately lead the way to sustainable development and also will be an attempt to reduce the global climatic crises.

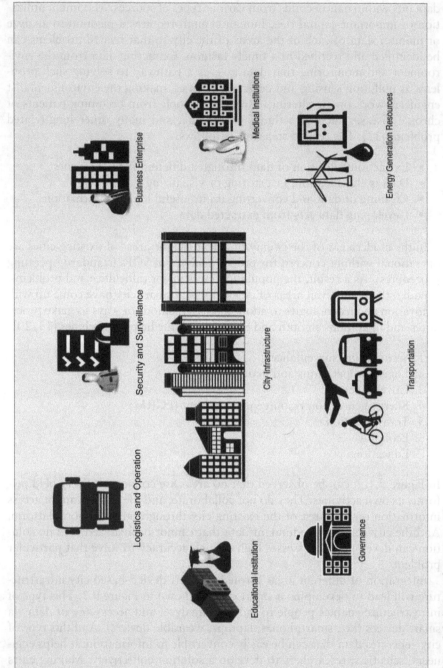

Figure 9.1 Areas of existing cities.

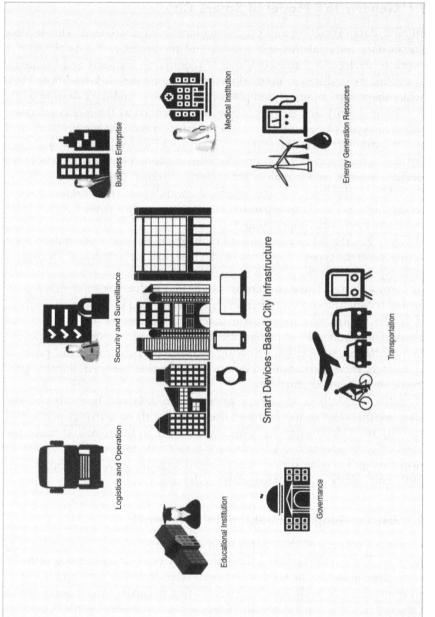

Figure 9.2 All areas integrated with smart devices.

9.3.1 Generic IoT Model of Smart City

Until now, data extraction and processing have been discovered. This section discusses data retrieval through sensors and the sharing of data through a network of devices. A framework that is required to transmit and exchange data in clusters to different networks needs a strong network backbone. IoT provides them with an integrated platform for devices and data sharing while maintaining security and privacy. Since IoT is defined as the interconnection of different types of networks with the Internet to increase connectivity and communication between different devices or machines to respond to a change of environment, it acts as a medium of communication between different types of devices, sensors, and machines through the Internet. Due to advancements in the field of technology, an increase efficiency and speed of communication is necessary. The SC has been separated into distinct networks with a particular generation of networks supporting it.

In Figure 9.3, the distinct network of each area in the smart city is developed by using different types of sensors or devices that are integrated to form a specific type of network, as shown. Together, these different types of the network form a cluster. These clusters are connected with common data management and computing centers. After segregation and data processing from these centers, the information moves to management centers for distribution and analysis. The major role of management centers is information management between different authorities, involving access to particular strata of peers. It also manages the information access and modification between different government and private organizations.

Scientists, researchers, engineers work on the data and innovations. The analyst use that data to monitor and keep track of those activities performed in the city on a day-to-day basis and work on it to improve and make the process efficient and faster. IoT places a major role in the centralization of information and making it more accessible to humans, which develops a user-friendly and collaborative environment in the city.

9.3.2 Establishing 6G Enabled–IoT Network

The primary sources of data generation and extraction are living/nonliving organisms. The data is retrieved through the activities or monitoring done by sensors, depending on the type of data extracted from the data and management centers or sent to the user for visualization purposes. An example of this is a smartwatch that shows the individual the number of steps walked, calories burnt, and pulse rate monitoring, all shown on the watch's display. Another example of data processing and management is traffic monitoring systems, the feeds captured by CCTV (closed-circuit television) cameras to manage traffic and accident investigation. This type of data is sent to the data management and computing centers.

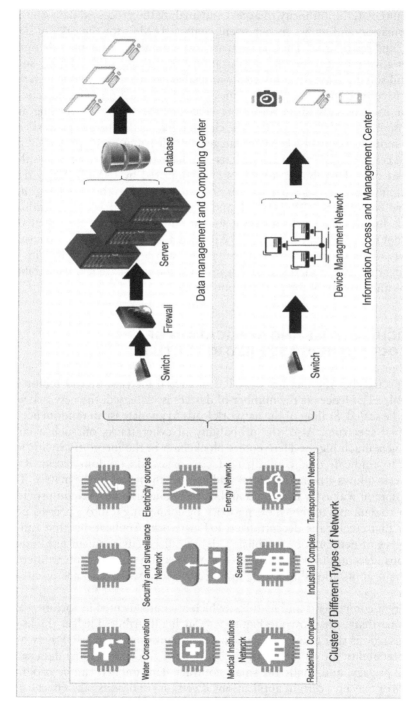

Figure 9.3 Generic IoT model of smart city.

In Figure 9.4, the authority to access and analyze the processed data is given to information and access management centers (IAMC) for the sake of data security and restrictions. These centers are further authorized to provide rights to other government or private organizations for the sake of developing solutions and solving real-world problems or challenges. Every component in this system is connected with a wireless mesh that is enabled on 5G connectivity to make each process faster and more efficient in terms of networking and distribution. Since data is stored in the cloud, it makes the storage process easier and more manageable. IAMC manages data distribution by providing the rights to certain registered organizations who have the authority to access that particular data. Distributed networks are monitored by IAMCs. Every data generation source acts as a participant node in a city when they are integrated to form a peer-2-peer (P2P) network and further developed as a decentralized network. But IAMC's are kept over it so that no node becomes corrupted and data is not tampered with by any of the participants in the network. To ensure the safety of data transactions and exchanges, each organization maintains its own ledger of its transactions, which upholds the transparency, trustability, security, and privacy of the organization's data.

9.4 ARCHITECTURE AND APPLICATION OF BLOCKCHAIN IN ASSIMILATION WITH IoT

It is expected that the number of IoT devices will increase to 130 billion by 2025 [14, 21]. Hence, as the number of devices is increased, the network will have to be scaled. Scaling of the network leads to an increase in random access and signal spectrum. Also, the probability of cyberattacks on such a large system gets much higher. Here enters blockchain in the integration with IoT devices to eradicate one jeopardizing factor of attack: random access. Random access allows attackers to enter and hamper the nodes of the matrix. The most common way is by data theft and its malicious use. Blockchain provides security to data and to the whole network by connecting it into a peer-to-peer manner that creates the decentralized IoT network. Applications that run in these types of networks or establish a decentralized environment are known as DApps (decentralized applications). In this section of the chapter, different blockchain applications and their architecture in a smart city are discussed further.

Different components and features, which are implemented in various areas of the smart city, are shown in Figure 9.5. Such a distributed ledger tracks all the transactions that take place among the organization or in a P2P network, data accessibility for all the participating nodes while maintaining data security and privacy, and finally the smart contract. This implicit calling program controls or executes certain applications if certain conditions take effect.

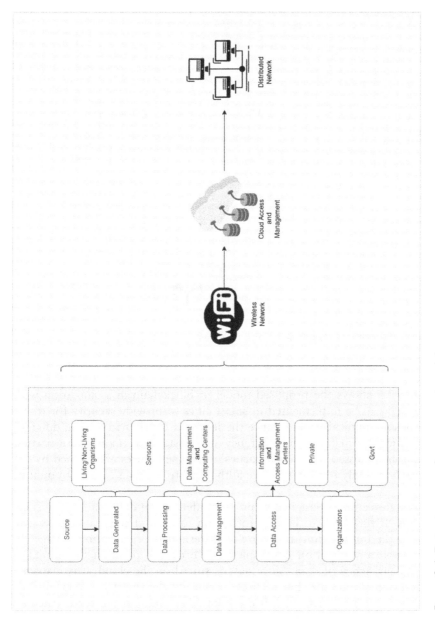

Figure 9.4 Data storage and management model.

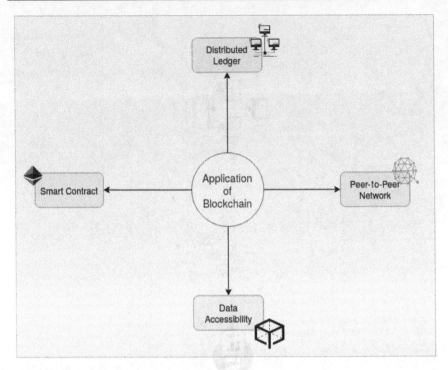

Figure 9.5 Blockchain components and features.

Figure 9.6 shows the proposed model of blockchain in assimilation with IoT, which can be implemented in smart cities to provide security for transacted or communicable data. Since the ledger is distributed among different organizations to maintain transparency and trustability among third parties, data can be accessed only when parties' digital signatures are verified by the organization. Only third parties are authorized to access a certain amount of data.

This architecture is designed to provide safety when data reaches the stage of access. It turns out to be in information form, and smart contracts give access to the information to private or government organizations. After all, the process of verification is completed in order to maintain the granularity and consistency of data. Since data is stored on the cloud, it is managed and accessed cloud-only, but a check is kept by the smart contract until or unless its condition is satisfied. It does not allow the party to read/write any form of data through the database. When IoT sensors and networks are connected at an intermesh level, it requires more safety, so that tampering in any form should not take place. Smart contracts are designed to make this process autonomous and effective.

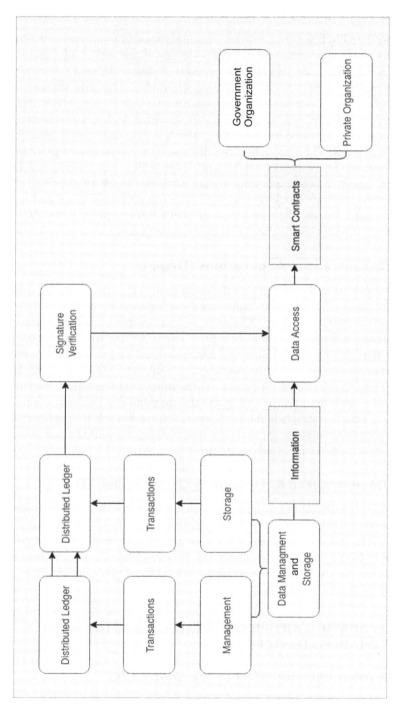

Figure 9.6 Architecture of blockchain in integration with IoT devices.

9.4.1 Decentralized IoT Network

A decentralized IoT network provides a scalability and tamperproof P2P network of devices that connects them in a decentralized way. Only connected and participant node devices are authorized to produce and share data, which restricts random access occurring among the nodes. The data sharing took place in this type of matrix in a transactional manner (i.e., when data is transferred from one node to another, it is recorded with a timestamp). The transactions are recorded in one ledger, which is also distributed among the authorized nodes. The transaction is validated by the miners and placed into block formats. These blocks are encapsulated with a timestamp, and hash values are generated corresponding to that particular block by a hashing algorithm. Every previous block is connected with the successor block and forms the blockchain. During the transaction of data from one point to another, the blockchain provides security and makes it tamperproof.

9.4.2 Decentralized Applications (DApps)

Web3.0 introduces the concept of decentralized applications, which run on smart contracts [15, 22]. Smart contracts are self-executing programs that run based on a certain agreement. The underlying principle of smart contracts is based upon the Turing machine concept. These contracts are also responsible for creating a decentralized network. Decentralized organizations run based on smart contracts, and those organizations are known as DAO (decentralized application organizations). A group of developers manages these organizations, and code is maintained and updated autonomously by smart contracts. Until now, smart contracts had limited applications. But when integrated with IoT, they have numerous applications: maintaining trustability and security among trusted and nontrusted organizations.

9.4.3 Decentralized Autonomous Organization (DAO)

This is an institutional organization that developers run. It is also considered an autonomous organization where no central authority is required for management. The rules of these organizations are written in the form of codes, which are also known as smart contracts. It is an online running technology. Smart contracts mostly run on ethereal networks. These organizations are managed by the P2P network [16, 23].

9.5 INTEGRATING ARTIFICIAL INTELLIGENCE (AI) WITH IoT-ENABLED DEVICES

9.5.1 Implementation of AI in the Smart City

The term artificial intelligence" came into recognition at a conference of Dartmouth College in the United States in 1956 [17–24]. Researchers started focusing on the great variety of its applications from language simulations to learning machines. Artificial intelligence is completely data driven. For scientists, it was a source of curiosity and skepticism, albeit in different measures. After six decades of its discovery, it is no longer confined to labs and has come into existence because of technological advancement. The other factors contributing to its recent progress are increased computing capacity, which is required to train larger and more complex models, to aid development, and to increment the efficiency of graphics and tensor processor units. Hyperscale clusters have added capacity, providing access to users through the cloud [18, 25].

One of the key factors is the massive amounts of data being generated in smart cities that can train the AI algorithms. An example of this is the autonomous vehicle, which come into existence because of innovation in sensors, LIDAR, mapping and satellite technology, navigation algorithms, and computer vision—all brought together and formed into an integrated system.

9.5.2 Applications of AI in the Smart City

IoT plays a significant part in most smart city applications and is responsible for generating a massive amount of data. When it comes to handling and manipulating extracted data, it gets strenuous to decide what action is precisely accurate or efficient. Analysis of the enormous data being accumulated through advanced techniques such as machine learning (ML), deep learning (DL), artificial intelligence (AI) to achieve an optimal solution.

In projects like smart cities, sectors can be observed where AI is playing a vital role, such as intelligent transportation system (ITS), physical cybersecurity, microgrid (MG), next-generation communication (5G). To enhance productivity and make models scalable, it is required to make effectual use of AI, DL, and ML techniques to manage big data analytics.

9.5.2.1 Intelligent Transportation System (ITS)

Machine learning– and deep learning (DL)–based techniques are playing a significant role in the development of the modern intelligent transportation system (ITS) such as in autonomous cars. It ensures the safety of passengers, lane, and edge detection [19, 26]. Protocols have been designed for such types of vehicles based on DL techniques. Veres et al. and others [26, 33] explained a detailed study on the role of DRL and ML to address several issues such as accident detection, traffic management, fleet management, etc. in a smart city that can contribute to enhancing the efficiency of processes in ITS. The authors developed a study of deep reinforcement learning (DRL)–based techniques, which throw light such issues as the design of trajectory, management of fleets, and physical cybersafety, which can play a major role in the development of the city.

9.5.2.2 Cybersecurity

Another important sector is physical cybersecurity, which is the most important aspect of the smart city and which can be realized as a significant smart city concept. Sensors, relays, and actuators are integrated for the purpose of gathering and transmitting data, a process that must occur safely and securely. This type of interconnectivity of devices can initiate or call to cybersecurity issues that need to be eliminated [22–29]. The authors of a number of works have proposed a security model for IoT devices in a smart city, machine learning–based architecture called the anomaly detection–IoT (AD-IoT) system [23–30]. Any suspicious activity happening at fog nodes can be detected with this technique by utilization of ML-based data set evaluation.

The authors proposed a model of deep reinforcement learning–based architecture for safeguarding the virtual infrastructure of a smart city from cyberattacks. The proposed model predicts the intruders at early stages based on data search behavior, and the cluster of the network can be safeguarded from such types of attacks.

9.5.2.3 Microcrid (MG)

In SCs, microgrids' operational structure is revolutionized by big data and focuses on enhancing its production efficiency. MGs are great examples of modern ICT systems, integration of IoT devices, and enormous data [25–32].

Operational and management decisions can be taken after analysis of data that arrives from different parts of the SCs. By the observation of the current scenario of MGs, the grids can be identified that are playing a vital role in effectively utilizing smart meters (SM) in big data by assessing load and predicting, estimating baseline, clustering load, detecting malicious data attacks. Big data analysis can be accomplished through phase measurement units (PMUs), which are generally used for estimating states, calibrating the dynamic model, grid visualization of transmission [33, 34].

9.6 COMPARISON OF DIFFERENT GENERATIONS OF COMMUNICATION NETWORK

9.6.1 First Generation (1G)

The very first generation of commercial cellular networks was introduced in the late 1970s and was commercialized in 1987 by Telecomm (known as Telstra). First generation is an analogue technology; the phones generally had poor battery life, and voice quality was largely without much security and would experience dropped calls. The maximum speed was 2.4 GHz. The problems of this generation are bad voice quality, poor battery consumption, and big cell phones [33–35].

9.6.2 Second Generation (2G)

1G used analog radio signals. Introduced in 1991, the second generation arrived with an improvement over 1G: It used digital radio signals. The motive for developing this generation was to have a secure and reliable communication channel. Implementing the concepts of CDMA and GSM provided small data services like SMS and MMS. 2G provided the capabilities achieved by allowing multiple users on a single channel via multiplexing. Fundamental services introduced by 2G were SMS, internal roaming, conference calls, call hold. It was better than the first generation since it allowed text message services. Whether the signals are weak or strong doesn't matter; text messages are not distorted.

9.6.3 Third Generation (3G)

Wireless technology was introduced by 3G in 2000. It used a new technology called universal mobile telecommunication system (UMTS) as its core network architecture, based on a set of standards for mobile devices and mobile telecommunication, the 2000(IMT-2000) specification by the International Telecommunication Union (IMT). One of the requirements set by IMT-2000 was that the speed must be at least 200 Kbps to call it a 3G service. It has multimedia services support along with streaming and offers the capability of universal access and portability across different device types. Also, it increased the efficiency of the frequency spectrum by improving compressed audio during calls so that more simultaneous calls can run in a frequency. Services introduced by this generation are video calling, video conferencing, live television. The devices that are compatible with this network are known as smartphones.

9.6.4 Fourth Generation (4G)

4G was developed to provide high speed, quantity, and capacity to users while improving security and lowering the cost of voice, data services, multimedia, and the Internet over it. Key technologies that have made this possible are MIMO (multiple input/multiple output) and OFDM (orthogonal frequency division multiplexing). Two important 4G standards are WiFi-Max and LTE. The max speed of a 4G network when the device is moving is 100 MBPS. It is 1 Gbps for low-mobility communication, such as when the device is stationary or moving at a walking pace. Latency is reduced by 300 ms to less than 100 ms with significantly lower congestion. Significant characteristics of 1G are mobile multimedia, anytime anywhere, global mobile support, integrated wireless solution, and customized personal service (MAGIC).

9.6.5 Fifth Generation (5G)

5G was developed to improve on the 4G network with higher data rates, connection intensity, much lower latency, and many other improvements. Plans for 5G included device-to-device (D2D) communication, low battery consumption, improved wireless convergence. The maximum speed of 5G was intended to be as high as 35.46 Gbps, which is over 35 times faster than that of 4G. Major technologies on which this generation was based on massive MIMO [33]. Different estimations have been made for the date of the commercial introduction of 5G networks. Next Generation Mobile Networks Alliance feels that 5G should be rolled out by 2023 to meet business and consumer demands.

9.6.6 Sixth Generation (6G)

Forthcoming 6G networks will be capable of integrating intelligent 5G and next-generation IoT infrastructure with advanced edge computing hardware support, which will also support the execution of heavy computational resources. This generation is considered a planned successor of 5G and is predicted to be faster from 5G at a speed of ~95 Gbit/s. 6G networks will be broadband cellular networks, which are estimated to be divided according to geographical areas, or cells. In the 6G mobile network, it is that an expected billion EDGE-based devices can connect and be capable enough to execute AI. Other types of algorithms (e.g., blockchain for security) over an SDN/NFV software control and management platform should be seen as its "brain." The evolution of this generation will lead to a proliferation of smart and wearable devices that will make IoE (Internet of Everything) a reality. The launch of the commercial introduction of 6G is expected by 2030.

9.7 CASE STUDIES

9.7.1 India's Smart Cities Mission

The Smart Cities Mission initiative is aimed at providing smart solutions to the problem of urbanization; the main objective of this mission is to follow a bottom-up approach in the case of town planning, providing basic and core infrastructure and decent and sustainable environment. Researchers had estimated that, by 2050, 50% of India's population will have migrated to cities and that, to accommodate them, 500 smart cities are required. To resolve such problems, two major smart city programs have been initiated recently: the Providing Urban Amenities to Rural Areas (PURA) mission and the Rural–Urban (RURBAN) mission [36]. These programs have been initiated to overcome the pressure on cities and to provide new avenues to village people. Under this mission, it has been estimated that 300 villages across the country will be developed as growth centers for the area by creating city-like infrastructures.

This will not be limited to only infrastructure development, but also the main focus will be to enhance digital and physical connectivity, as well as quality healthcare and educational infrastructure. The main aim of these villages will be to provide skill development, digital literacy, quality sanitation, an agroprocessing industry, among others. Therefore, the Rurban mission will provide a better quality of life and increased urban opportunities in the villages themselves.

9.7.2 Birmingham Digital Revolution

The roadmap of Birmingham's smart city was launched in 2014 in association with Digital Birmingham (DB), a council-owned partnership-based alliance organization. This road map was based on the categorization of three major themes: human resource, technology, and infrastructure, economy. They focused on digital inclusion, improving citizens' ICT skills, and business operating processes in human resources. The area of technology and infrastructure requires major focus given their chief aim to enhance broadband connectivity and make open data accessible. Next, in the economy, a major concern was automating the process of social care and development in energy regulatory resources and smart mobility through which they want to build an economically feasible solution. This road map was dedicated to existing projects of DB, such as investment over ultrafast broadband and various other initiatives based on developing technological skills. £26 million was invested through the Urban Traffic Control Major Scheme (UTCMS) to integrate transport data from different agencies to a simple platform. It also participated in the European Union–funded initiatives for smart space projects [35].

9.7.3 Smart City Bristol Project

This project was launched in 2011 with the aim to use smart technology to meet its motive of reducing CO_2 emissions by 45% by 2020. The team of councils was mostly focused on smart transport, energy, and data. This initiative also includes some funded pilot projects which the European Union funded, such as Smart Spaces. In 2013, Bristol also granted £3 million for opening the City Living Lab from TSB Bank for integrating data from different resources to host events to encourage students and businesses to use them. Recently, the council focused on implementing many challenges for citizens' feedback through Bristols Open Data Energy Challenge and the George Idea Lab [37].

9.8 SMART CITY CHALLENGES AND OPPORTUNITIES

Challenges have to be faced by private and government organizations to implement modern technology in integration with artificial intelligence. Given the scope and range of its applications forecast in industry, use of AI become a major requirement to meet the demands. The challenges are categorized and discussed in this section.

9.8.1 Economic Challenge

Implementation of AI requires a financial investment by an organization. Mass implementation could have a great economic impact on the institution, so affordability and maintenance of computational expenses will play a major role in this challenge. Since AI's accuracy and precision are increased through the high volume of captured or real-time feed data, managing its storage also requires capital investment. Execution of AI depends on the sector or enterprise.

9.8.2 Social Challenge

The rise in the use of AI is likely to oppose cultural norms and can act as a potential barrier for some of the sectors. For example, in healthcare, major changes will occur, such as in interactions with the patient. The impact on the clinician and patient will be significant. Arguably, the adoption of AI in radiology can be considered a major barrier. Also, patients should possess some knowledge about the interaction with such technologies. This challenge can act as a major potential obstacle in the way of adoption of AI.

9.8.3 Managerial and Institutional Challenges

Adoption of AI leads to a lot of transitional challenges, which can have strategic implications. The measure of success can be defined based on financial return in terms of investment and trust. AI researchers should focus more on the improvement in humans and computer interaction. It should be connected with a flow of information. Lack of focus on strategic implications can turn into a major cause of failures in operations or standard operating procedures (SOP) of an enterprise. AI researchers should understand the need of the present scenario in healthcare systems and design technologies to solve the problems of that sector.

9.8.4 Ethical Challenges

In the current scenario, each dimension should be analyzed before the implementation of AI to utilize it to the maximum extent. Focus should be directed on the possibility that there can be negligence in adopting this technology either on an organizational level or on an individual basis due to concerns related to the ethical dimensions of such systems. The immediate forecast in the development of AI technologies can raise the issue of ignoring ethical issues, which formally highlights the need for building such policies and regulations and bringing new reforms to prevent the misuse of AI on ethical grounds.

9.8.5 Future Opportunities for AI

Studies show that, by 2030, 70% of enterprises or industry sectors will implement some form of artificial intelligence in their processes or operations. It can be expected that future factories or industries will be likely to adopt AI technology to automate their production processes. This will lead businesses to migrate toward intelligent platforms developed by the integration of AI and cyberphysical capabilities. In the healthcare industries, AI will open the doors of new opportunities for the application of AI, such as in medical diagnostics and pathology, where tasks can be precisely, efficiently, and quickly automated. Health monitoring can occur through sensors that can be part of bio–field technology, by which, studies suggest, numerous life sign parameters can be monitored through body area networks (BANs), whereas diagnosis will require a clinical opinion and arbitration by humans. The incorporation of AI with marketing also generates customer requirements and profiles. When information on customer requirements and demographics is collected by analyzing big data on that basis, purchasing habits can be predicted. Understanding consumer needs in the supply chain becomes a critical factor that can be resolved by AI, which can act as a censorious element.

9.9 CONCLUSION

Through the study of smart cities and their application, greater advancement in technological terms will make these cities smarter. Analysis has shown that different applications require different technical backbone supports. But IoT provides them with a single platform where everything is emerged, and it can be useful for multiple applications at a time. A brief study of artificial intelligence gives an overview of process automation that can revolutionize the industrial sector. Blockchain can handle the secured transmission and distribution of data and thus enables us to build a trustless and open network for recording data transactions in the P2P network. This chapter has briefly described the application of modern technology in integrated environments, which can be useful for multiple applications in a smart city. In this chapter blockchain-based, IoT architecture has also been mentioned, which helps develop the trust-less network of peers.

REFERENCES

1. Su, K., Li, J., and Fu, H. *Smart City and the Applications*. School of Computer, Wuhan University (2011).
2. Madakam, S., and Ramaswamy, R. 100 New Smart Cities (India's Smart Vision). 5th National Symposium on Information Technology: Towards New Smart World (NSITNSW) (2015).

3. Mulligan, C.E.A., and Olsson, M. Architectural Implications of Smart City Business Models: An Evolutionary Perspective. *IEEE Communications Magazine*, 51 (6), 80–85 (2013).
4. Arroub, A., et al. A Literature Review on Smart Cities: Paradigms, Opportunities and Open Problems. 2016 International Conference on Wireless Networks and Mobile Communications (WINCOM) (2016).
5. Saravanan, K., Golden Julie, E., and Harold Robinson, Y. Smart Cities & IoT: Evolution of Applications, Architectures & Technologies, Present Scenarios & a Future Dream. In *Internet of Things and Big Data Analytics for Smart Generation*. Springer, 135–151 (2019).
6. Wenge, R., et al. Smart City Architecture: A Technology Guide for Implementation and Design Challenges. *China Communications*, 11, 56–69 (2014).
7. Arasteh, H., et al. Iot-Based Smart Cities: A Survey. 2016 IEEE 16th International Conference on Environment and Electrical Engineering (EEEIC), IEEE (2016).
8. Chakrabarty, S., and Daniel Engels, W. A Secure IoT Architecture for Smart Cities. 13th IEEE Annual Consumer Communications & Networking Conference (CCNC), (2016).
9. Impedovo, D., and Pirlo, G. Artificial Intelligence Applications to Smart City and Smart Enterprise. *Google Scholar*, 2944 (2020).
10. Khan, W.P., Byun, Y.C., and Park, N. A Data Verification System for CCTV Surveillance Cameras Using Blockchain Technology in Smart Cities. *Electronics*, 9 (3), 484 (2020).
11. Tanwar, S., et al. Machine Learning Adoption in Blockchain-Based Smart Applications: The Challenges, and a Way Forward. *IEEE Access*, 8, 474–488 (2019).
12. Gupta, A., and Jha, R.K. A Survey of 5G Network: Architecture and Emerging Technologies. *IEEE Access*, 3, 1206–1232 (2015).
13. Lu, H.P., Chen, C.S., and Yu, H. Technology Roadmap for Building a Smart City: An Exploring Study on Methodology. *Future Generation Computer Systems*, 97, 727–742 (2019).
14. Buterin, V. A Next-Generation Smart Contract and Decentralized Application Platform. *White Paper*, 3, 37 (2014).
15. www.idc.com/getdoc.jsp?containerId=prUS45213219.
16. Li, J., Wu, J., and Chen, L. Block-Secure: Blockchain Based Scheme for Secure P2P Cloud Storage. *Information Sciences*, 465, 219–231 (2018).
17. www.mckinsey.com/featured-insights/artificial-intelligence/the-promise-and-challenge-of-the-age-of-artificial-intelligence.
18. Veres, M., and Moussa, M. Deep Learning for Intelligent Transportation Systems: A Survey of Emerging Trends. *IEEE Transactions on Intelligent Transportation Systems*, 21 (2019).
19. Ferdowsi, A., Challita, U., and Saad, W. Deep Learning for Reliable Mobile Edge Analytics in Intelligent Transportation Systems: An Overview. *IEEE Vehicular Technology Magazine*, 14 (1), 62–70 (2019).
20. Rawat, D.B., and Ghafoor, K.Z. *Smart Cities Cybersecurity and Privacy*. Elsevier, 2018.
21. Alrashdi, I., Alqazzaz, A., Aloufi, E., Alharthi, R., Zohdy, M., and Ming, H. AD-IoT: Anomaly Detection of IoT Cyberattacks in the Smart City Using Machine Learning. 2019 IEEE 9th Annual Computing and Communication Workshop and The Conference, CCWC, 0305–0310 (2019).

22. Elsaeidy, A., Elgendi, I., Munasinghe, K.S., Sharma, D., and Jamalipour, A. A Smart City Cybersecurity Platform for Narrowband Networks. 2017 27th International Telecommunication Networks and Applications Conference, ITNAC, IEEE, 1–6 (2017).

23. Shahinzadeh, H., Moradi, J., Gharehpetian, G.B., Nafisi, H., and Abedi, M. Iot Architecture for Smart Grids. 2019 International Conference on Protection and Automation of Power System, IPAPS, IEEE, 22–30 (2019).

24. Shahinzadeh, H., Moradi, J., Gharehpetian, G.B., and Mehrdad Abedi, H.N. IoT Architecture for Smart Grids. 2019 International Conference on Protection and Automation of Power System, IPAPS, 22–30 (2019).

25. Bhattarai, B.P., Paudyal, S., Luo, Y., Mohanpurkar, M., Cheung, K., Tonkoski, R., Myers, K.S., Zhang, R., and Zhao, P. Big Data Analytics in Smart Grids: State-of-the-Art, Challenges, Opportunities, and Future Directions. *IET Smart Grid*, 2 (2), 141–154 (2019).

26. Karimipour, H., Geris, S., Dehghantanha, A., and Leung, H. Intelligent Anomaly Detection for Large-Scale Smart Grids. 2019 IEEE Canadian Conference of Electrical and Computer Engineering, CCECE, IEEE, 1–4 (2019).

27. Du, D., Chen, R., Li, X., Wu, L., Zhou, P., and Fei, M. Malicious Data Deception Attacks Against Power Systems: A New Case and Its Detection Method. *Transactions of the Institute of Measurement and Control*, 41 (6), 1590–1599 (2019).

28. Wang, Y., Chen, Q., Hong, T., and Kang, C. Review of Smart Meter Data Analytics: Applications, Methodologies, and Challenges. *The IEEE Transactions on Smart Grid*, 10 (3), 3125–3148 (2018).

29. Fallah, S.N., Deo, R.C., Shojafar, M., Conti, M., and Shamshirband, S. Computational Intelligence Approaches for Energy Load Forecasting in Smart Energy Management Grids: State of the Art, Future Challenges, and Research Directions. *Energies*, 11 (3), 596 (2018).

30. Tureczek, A., Nielsen, P., and Madsen, H. Electricity Consumption Clustering Using Smart Meter Data. *Energies*, 11 (4), 859 (2018).

31. Harkut, D.G., and Kasat, K. *Introductory Chapter: Artificial Intelligence-Challenges and Applications*. Artificial Intelligence—Scope and Limitations (2019).

32. http://net-informations.com/q/diff/generations.html.

33. https://rantcell.com/comparison-of-2g-3g-4g-5g.html.

34. Shyama Prasad Mukherji Rurban Mission (SPMRM). *Ministry of Rural Development Government of India*. https://rurban.gov.in.

35. Digital BirMingham. *Information Technology and Digital Service*. http://digital-birmingham.co.uk/about/.

36. Liu, P., and Peng, Z. China's Smart City Pilots: A Progress Report. *Computer*, 47 (10), 72–81 (2013).

37. Bristol Connected City. *Connecting Bristols* https://www.connectingbristol.org/wp-content/uploads/2019/09/Connecting_Bristol_300819_WEB.pdf

Chapter 10

Blockchain

Insights Related to Application and Challenges in the Energy Sector

Iram Naim,[1] Ankur Gangwar,[2] Ashraf Rahman Idrisi,[3] and Pankaj[4]

CONTENTS

[1,2,4] Department of Computer Science and Information Technology, MJP Rohilkhand University, Bareilly, India
[3] Department of Maintenances and Services, Bharat Heavy Electrical Limited, New Delhi, India
Corresponding author: Iram.naim03cs@gmail.com

DOI: 10.1201/9781003264392-10

10.1 INTRODUCTION

The sixth-generation (6G) network with a better transmission rate and reliability than the older network are becoming more popular nowadays. A blockchain is an efficient approach with great significance to the 6G network and other networks. A combination of blockchain with 6G will enable the network to monitor and manage resource utilization and sharing efficiently in many domains. A blend of power, electronic, and information technologies with intelligent management techniques for energy sharing is considered the energy Internet [1]. The conventional trading technique in the field of energy and its structures is based on central power system models. But the latest technologies have provided the freedom to trade energy in the open market and have facilitated the segregated renewable and clean energy to compete in the open market [2]. The energy sector is changing rapidly and is focused on accommodating embedded renewable energy sources. Renewable energy is today's need and is prioritizing various industrial segments like transportation, cooling, heating, etc. to overcome the use of traditional fossil fuels [3]. Renewable energy sources have changed the market scenario, and the world is shifting its priority away from traditional fuel. There are many recent advancements in renewable energy sources due to the unbundling of the energy sector, privatization, financial developments, and energy policy [4]. To strengthen and enhance the share of renewable energy, a budget of \$12.9 trillion was provided to execute the weather trade and get renewable energy in line with the sector's electricity outlook 2017 record [5]. Funding for renewables will reach as much as \$12.9 trillion to support the tasks in coping with climate and generally to gain entry to the use of climate-friendly electricity. Even though renewable energy has a large capacity, its development pace is slow as it brings with it concerns like huge capital investment, technical feasibility, and reliability. Such indirect parameters are of great concern for investors.

Although renewable energy has given much strength to grid capacity, there is still wide scope for real-life customer adoption. Suppose electricity transmission and distribution become fully linked. In that case, customers can start purchasing electricity quickly from the grid with or without photovoltaic structures and using it. Researchers in renewable energy are looking toward a powerful tool, blockchain, to decentralize electricity so as to completely change and stabilize the transition phase of the grid [6].

Blockchain technology is advancing with time, whereas technologies like smart contract management and distributed ledger are also advancing moderately [7]. New framework agreements, along with technocommercial aspects, are emerging for blockchain techniques. However, the regulations for distributed the trading of energy and associated mechanisms like microgrids are already being used. Blockchain has an extra edge in trading in the energy market, the financial sector, and supply chain management of the energy sector [4]. The potential of blockchain can completely transform the energy sector. It is an advanced and trustworthy tool for financial services, energy services, and other vital services. Blockchain contains smart contracts, and its ease of operability will set a new direction to the energy sector. Its smooth management system will allow numerous devices to interact with one another. It will also pave a new path for electric vehicles, solar energy, and smart metering. Expansion in the interconnection and digitalization of systems has raised a concern for security. The use of blockchain can deal effectively with these advancements in the energy sector. The requirement to improve process proficiency, adaptability, and lead time needs innovation based on blockchain. In the future, the business scenario is due for a change due to the flood of electric vehicles, a thriving number of smart machines, and other advancements in technology.

10.1.1 Motivation for Blockchain Usage in the Energy Sector

The energy sector is undergoing massive advancement and changes to cope with the uncertainty in energy resources and variations in energy consumption. Renewable energy resources are being preferred worldwide over traditional energy sources. This shift of paradigm is also one of the key driving forces for energy sector industries.

The energy sector currently faces many changes in energy trading, digitalization, the Internet of things and transparency, imbalance settlement, smart grid applications, etc. There is a need for a secure mechanism in the energy sector to provide efficiency and better utilization of energy resources. Blockchain can establish a secure and decentralized resource-sharing environment [8]. This motivates the use of blockchain in various applications of the energy sector. Network-enabled blockchain can be utilized in the 6G wireless network for decreasing the administration cost required for dynamic access systems that improve spectral efficiency [8]. Blockchain supports the growth of 6G by mitigating several security threats related to, for example, spectrum and content sharing. However, blockchain usage is not without its challenges, especially about user privacy [9]. 6G blockchain can also be applied in taxonomy, field trials, and their challenges and opportunities [10].

10.1.2 Contributions of This Chapter

This chapter describes and discusses the current developments in the energy sector, various applications of blockchain in that sector, and details about challenges while acquiring blockchain. Although blockchain has been started in the energy sector, there is tremendous scope in which the blockchain can transform the outlook of the energy sector.

- This chapter highlights the attributes of blockchain technology in the energy domain that go beyond its basic use. The chapter will first dive into the use of blockchain in application areas of the energy domain that mainly cover bills for energy, microgrids, sales and marketing, trading, automation, smart grid applications, grid management, energy security, utilization of renewable energy sources, resource sharing, competition, digitalization, the Internet of Things, and the transparency, imbalance settlement.
- The role of blockchain in paving the path for turning these challenges into opportunities is also a major focus of this chapter. The challenges are always considered a bedrock of opportunities; this chapter will also highlight the opportunities for improvement in the energy sector by adopting blockchain in many potential areas like development cost, migration toward new technology, coordination with regulatory bodies, lack of flexibility, etc.
- The chapter defines blockchain utilization in the energy domain to better address performance achieved by the sixth-generation (6G) network.

10.1.3 Chapter Organization

The chapter first provides an overview of blockchain in Section 2. In Section 3, the applications of blockchain in the energy sector are discussed. Section 4 includes a discussion on several applications and challenges in the energy sector. Further, in Section 5, the chapter will also suggest the future scope of blockchain in this sector and the conclusion.

10.2 OVERVIEW OF 6G-ENABLED BLOCKCHAIN

Blockchain advancement is portrayed as a decentralized, scattered record-keeping system that records details in electronic or computer-based assets. As the name suggests, blockchain at the fundamental level is a chain of blocks (squares) [11]. Exactly when we express the words "square" and "chain" in this novel circumstance, we are truly describing cutting-edge advanced information (i.e., block) set aside in an open information or database (i.e., chain) [12].

In real-world blockchain applications, the connections are not that simple because the blockchain is a distributed technology rather than centralized. A block can be connected to thousands (or even more) blocks. Whenever a block is added to the blockchain at its farthest reaches, it is always difficult to modify the block's content. It is because each square consists of its own hash and the hash details of the previous block. It is well understood that hash

Table 10.1 Blockchain Performance Parameters 5G vs. 6G

Performance parameter	5G	6G
Data transfer speed	0.1–20 Gb/s	I Gb to I Tb/s
Possibility of error	10–5	10–9
Localization precision	10 cm in 2D	I cm in 3D
Density	106/km²	107/km²
Mobility	500 km/h	1000 km/h
Traffic capacity	10 Mb/s/m²	<10 Gb/s/m³
Latency	1–5 ms	10–100 ns

codes are created through a mathematical algorithm that converts electronic data into a series of alphabets and numerals [13]. In case the data is modified in any form, the hash code is also bound to change.

Blockchain helps in utilizing the Internet of Everything (IoE) for sixth-generation wireless network applications. It avoids the requirement of any type of diaries. 6G can use blockchain for many applications like access control, security, authenticity, etc. The objective of 6G is to meet the Internet of Everything requirements and enhance the technology. 6G has much improved key parameters over 5G, as provided in Table 10.1 [9].

10.3 BLOCKCHAIN APPLICATIONS IN THE ENERGY SECTOR

Many industrial segments have realized the potential of blockchain technology in different sectors. In the financial market, the blockchain has already proved its performance and developed trust. The energy sector is the major area where blockchain utilization can provide enormous change. Blockchain innovations in conjunction with IoT empower customers to buy energy directly from the network instead of from designated distributors. Blockchain can give buyers more noteworthy proficiency and power over the selection of their energy sources. Moreover, a changeless record gives secure and continuous updates of energy information. Blockchains guarantee an increasingly productive and versatile IT foundation for energy use [14]. An evaluation of blockchain in the energy sector is presented in Kumari et al. [15]. Further applications in which blockchain can perform a key role are discussed next.

10.3.1 Energy Bills

The use of smart contracts and advanced metering methods used in blockchain provides automated energy bills to the customers and energy producers [16].

Electricity generating organizations now benefit from small payment, paid as used, or prior payment methods [17].

These days, with the fast advancement of innovation and social awareness, prepayment plans are broadly accepted. There are many advantages to utilizing prepayment plans for customers and providers [18]. The month-to-month utilization costs can be monitored, the income of the suppliers can be improved on the basis of the prepayment nature before the item being used [19]. Presently, the estimation of power use is worked out through the old metering technique in unit terms of kilowatt per hour (kW/h), which depends on the power utilization for a specific period. To enable power sharing over the current power network framework, we have to change this sort of estimation technique onto another resource that is completely converted to computerized resources and moves over a distributed system. The way for changing electrical capacity to the computerized resources is called tokenization, and the tokenized resource is called Wattcoin. Wattcoin can be moved or constrained by Wattcoin wallet, which is a portable application that customers can install on their cell phones. Wattcoin use can be decreased in the power meter by altering the overall demand. Other than the decrease of the Wattcoins, the system additionally gives a continuous power use to individual or in industrial customer.

10.3.2 Microgrids

Microgrids can be defined as neighborhood power frameworks that can generate electricity and power on their own. These circulated grids offer sufficient energy to the utilities for security, flexibility, repetition, and backup capacity to protect against blackouts. Microgrids are dissemination-level systems interfacing different energy assets [20]. Most microgrids in the created world are joined to a bigger power network for stability. Microgrids can be modified to accommodate the power demand. They can be scaled to a discrete activity, for example, a business park, school grounds, social insurance complex, mine, military office, neighborhood, or basic city administrations. Microgrids are not an innovation; they have been secretly in other matrixes, which have given high-caliber, dependable power. Microgrids have been around for a considerable length of time in locations where there is no other option, i.e., in zones where there is no central electrical network or in zones temporarily connected with local and central grids.

As the market changes and sustainable power sources develop, blockchain offers a viable method to deal with the unpredictable and decentralized exchanges between clients, enormous and small scope makers, retailers, merchants, and utilities. Blockchain microgrid is a rising methodology, as explained in Brooklyn [21]. The innovation is increasingly similar to a shared and friendly showcase wherein an exchange between two agencies is encouraged. Additionally, it makes possible neighborhood-sized markets where nearby energy can be exchanged inside a particular virtual or physical microgrid [22].

10.3.3 Sales and Marketing

The sale of energy is directly linked with factors like a portfolio of customer, consumer inclination, and environmental factors [23]. Thus, it is important to identify the energy pattern for a consumer and provide customized energy services. Identifying such patterns can be easily obtained by combining blockchain and machine learning (a technique of artificial intelligence). With blockchain, direct energy sales from small producers to their neighbors can take place (generating fewer network losses); a P2P (peer-to-peer) proximity market could be created in which smart contracts manage negotiations automatically and (almost) in real time, taking into account supply and demand, available power, and other parameters monitored by smart devices [24]. AI algorithms could help propose or choose the best rates. Transactions might generate micropayments that can be made through custom cryptocurrencies or tokens.

Blockchain offers straightforwardness, for instance, in power shopper purchases (e.g., type and unit of power). This fulfills the expanding request from clients for simple and quick access to clear and understandable data about the merchandise and services they are buying.

10.3.4 Trading

Distributed trading models based on blockchain have the high potential to change various types of markets like management of the wholesale market. The models using blockchain can affect the energy market, its transactions in trading for green energy, and its certifications for which blockchain-based systems are also undergoing development. The electricity-generating companies release the power to the blockchain for trading purposes. In turn, the consumer, whether an individual or an organization consuming power, purchases the energy through the trading platform or directly from the selling company. Blockchain retains the transaction data related to buyer, seller, and energy tariffs, avoiding any transaction refusal. A secure energy trading mechanism for renewable energy resources between suppliers and consumers using cost-aware and store-aware algorithms has been proposed [25].

Power and energy trading organizations execute several exchanges every day to gain huge exchanging benefits. Organizations have contributed millions to work out different frameworks and to use computerized and manual procedures to overcome penalties. There is a valid justification for why the blockchain is important to such a significant number of money-related deals. The blockchain guarantees a value-based stage that is exceptionally secure, requires minimal effort, is quick, and has fewer instances of mistake, and offers the chance of lessening capital prerequisites [26]. While blockchain innovation is still in the moderately start-up phases of improvement, the potential uses are expansive and promising [27]. The research has proposed a smart contract–based secure energy trading scheme for a smart grid system for peer-to-peer energy trading [28]. Another research has presented a secure energy trading framework, having security and privacy preservation in demand response management [29].

The energy and product transaction life cycle, in any event, for basic transactions consists of many cycles within the company and with other representatives outside the company. The counterparties shall examine and adjust the transaction data at various points in execution through to the transaction settlement. Furthermore, due to the transaction life cycle, an organization may require collaboration with other counter-suppliers, traders, intermediaries, coordination suppliers, banks, controllers, and value columnists. As depicted in Figure 10.1, a web of information interfaces, frameworks, and cycles are needed to encourage such cooperation. Through the blockchain technique, there is greater possibility for the smooth conducting of interior and exterior activities.

10.3.5 Automation

Blockchains can execute the command of decentralized systems for energy and microgrids [18]. The adoption of local energy marketplaces is enabled by local P2P energy trading or distributed platforms that can significantly increase energy self-production and self-consumption, potentially affecting revenues and tariffs. Blockchain is a decentralized digital ledger regulated and monitored by a distributed network operating through prescribed protocols [30–31]. A convergence of various technologies belonging to network, data, consensus, identity, and automation management is essential for successfully creating and implementing a blockchain [32, 33].

Automation can be accomplished through PC programs called smart contracts, which are put on the blockchain and which characterize the authoritative commitments to exchange advantages between peers. Automation in blockchain is completed with the assistance of smart contracts that may characterize legally binding commitments, the care or movement of advanced resources, and the rights and benefits of hubs. These contracts are the utilization of blockchain innovation that will affect all product members. Smart contracts give more noteworthy automation and imitate activities that are mostly performed by outsiders or, on the other hand, by

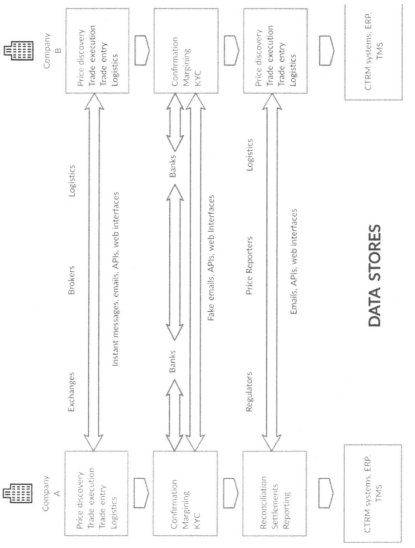

Figure 10.1 Energy and commodity transition life cycle.

middle people [34]. Smart contracts can further be divided into two major types [33]:

1. *Deterministic smart contracts*: These contracts are independent of any kind of information exchange from any outside agency. Such contracts are framed by utilizing the existing information available within the blockchain system.
2. *Nondeterministic smart contracts*: These contracts require data from outside agencies, for example, data related to weather. Such contracts are more flexible but are sensitive to attacks from outside.

10.3.6 Smart Grid Applications

Blockchains can significantly be utilized for sharing and storing information and communicating with smart devices [18]. The main components in the smart grid consist of smart meters, high accuracy sensors, control equipment, software, energy & building monitoring systems. The blockchain technique provides a reliable exchange of information to the smart grid and standardizes the systems.

Blockchain provides a major trading platform in smart grids [35]. Applying it removes third-party intervention and defines peer-to-peer trading of electricity between consumer and producer [36, 37]. The advantages of the blockchain-enabled trading system result in real-time market development, reduced transactional cost, and improved security to the user [38]. While using the blockchain to establish a more advanced and sharing network, computational capabilities could create complex applications. This provides a shared computing network that can be used to deliver a range of solutions within smart grids [39].

Most analysts in leading industries agree with the rise of blockchain. Theoretically, blockchain systems would follow advanced development and smooth change to the smart grid. Distributed technology also serves as a benchmark for some smart grid technologies. The cutting edge of blockchain and its benefits have created attraction in developing this technique in the smart grid. The main application of this technology in the field of blockchain can be divided into the following parts [40]:

- *Power production*: Blockchain infrastructure provides the suppliers with complete awareness in a real-time context of an electric grid's overall operation and control. This allows them to develop plans for delivery, leading to maximization of profits.
- *Power transition and delivery*: Blockchain networks allow the centers of computation and control to provide decentralized structures that can solve the core obstacles of conventional centralized systems.

- *Power utilization*: Blockchain may help handle the energy exchange between the producers and the various energy storage systems and electric vehicles.

10.3.7 Grid Management Energy Security

The blockchain technique helps manage distributed networks, assets, and flexible services. It can resolve the problem of the expensive network by providing combined flexible trading models and improved utilization of resources. Thus blockchain provides an effective tariff structure with networking [16]. The use of cryptographic technology improves the reliability of transactions and enhances the security of data. The additional advantages of blockchain are privacy, the security of data, and identity management.

The use of blockchain in smart grids may benefit our existing and future electrical power networks. Many of the benefits to the electrical power network directly relevant to the properties and operational standards of blockchain come from the decentralized infrastructure for trade development. The key benefits associated with the features and operating standards of blockchain within the electrical power structure are the decentralization of security, increased protection, increased durability, increased accountability, scalability, less complexity, and greater computational power.

Figure 10.2 explains why specific blockchains may be used to secure data in various smart grid layers. For this example, a separate blockchain aggregator

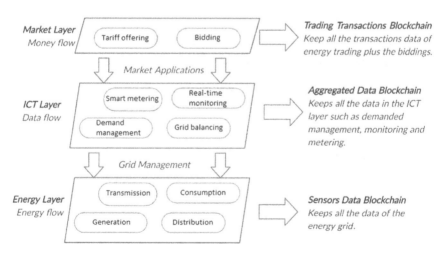

Figure 10.2 Blockchain applications framework in smart grid security and data protection.

manages each layer of the smart grid. Such blockchains shall have the means of securing the smart grid by offering a stable and safe network for data storage. In addition, there is no clear connection between the identification inside the blockchain system and identifying the individual consumer/producer. Therefore blockchain reduces the susceptibility to malicious attacks and increases the electrical system's durability [41].

10.3.8 Utilization of Renewable Energy Sources

Blockchain technology provides an unparalleled potential to support distributed solar resources, with the characteristics of decentralization, openness, flexibility, and stability. Blockchain technology's functionality and strengths follow the specifications of a mutual and credential industry. It allows direct interactions between producers and consumers in a costless manner without the need for intermediaries, which could lead to offline communications in communities and impact people's attitude. Because of its transparency and immutability, it is possible to construct a reliable record of certificate generation and transactions, making the system trustworthy and attractive. In smart contracts, effective compensation schemes may be programmed, implemented, and enforced automatically on blockchains. Blockchain technology is worth exploring to facilitate the businesses for energy certificates and distributed renewable energy adoption [42–44].

10.3.9 Resource Sharing

Blockchain can provide effective resource-sharing solutions among various consumers, such as the sharing of the setup created for charging electric vehicles [44], of information, or of nondistributed community storage.

Better performance can be achieved by the sixth-generation network over past generation networks as it tries to meet the demand of rising technologies and applications. On the other hand, the appropriate use of resources is hard to achieve. Using blockchain, resource management, and sharing among multiple users is easy in 6G networks [45]. It also serves to execute a shared economy in the energy sector, resulting in new business models and electricity democratization [46]. Dai et al. [19] has presented blockchain and deep reinforcement learning for efficient resource management services, including spectrum sharing and energy management.

10.3.10 Competition

Changing energy suppliers is now easily possible in a fast track mode using smart contracts [23]. The increase in mobilization in the market enhanced the competition and optimization in the prices of energy. In the traditional energy market, collecting the cost of electricity is an important and tidy process. In contrast to the traditional market, the blockchain system provides an easy realization for collecting such energy costs. In the traditional system, the collecting is performed in real time. However, in blockchain, smart contracts are used to provide the transaction details of the user to the respective suppliers.

Simultaneously, it is always desired to promote the act of producing and consuming green energy as per the guidelines of respective regulatory authorities and to create awareness for both consumers and energy-generating agencies. Blockchain also serves as an effective platform for the dissemination of such policies. Blockchain also considers the trading medium and records trading and keeps a note of green energy. Thus smart contracts are effectively used in blockchain to provide subsidies and other advantages to respective consumers and to reduce the processing required in traditional systems [33].

10.3.11 Digitalization

Blockchains facilitate digitalization using digital P2P transfers, so machine-to-machine (M2M) connectivity and data sharing between smart devices is theoretically feasible. Smart meters and ICT devices are gradually being used in control networks in the energy sector [47]. It is estimated that the number of smart meter readings alone will grow from 24 million per annum to 220 million a day for a big utility. Combined with the influence of automation and big data analytics, this development may eventually change the energy market supply chains. Useful information from the data will boost the efficiency of the power network and diagnostics of equipment and contribute to cost savings. Digitalization offers an incentive for electricity utilities to enhance network performance, payment procedures, and supply chain and to seek different avenues of competition and alternative business models. Data usage could contribute to optimizing market collection and demand response systems to promote virtual power plants (VPPs).

Digitalization has reached virtually all markets, producing innovative goods and services, different manufacturing methods, and new forms of communicating with consumers and other sectors [48]. Digitalization will more or less transform all markets, whether more or less slowly or quickly. Internet transformation has had a significant effect on the role that information and communication technology (ICT) plays in virtually every sector. And ICT and the Internet of Things (IoT) play a key role in the transformation process of the energy industry. Digitalization allows for new market participants to interfere with the solar industry as well. Competition is getting intense, and new entrants are moving into the market. But digitalization also offers energy members the opportunity of improving processes, new ways of reaching and interacting with customers, changing how they are partnering and who they are partnering with, and it offers the opportunity to generate commercial opportunities by developing new products or services and innovative business models. New platforms and business shapes will emerge, and even some new platform-based business systems in collaboration with various other players in the market will emerge due to digitalization.

Digitalization affects or at least can have a very strong effect on certain industries, user preferences, consumer desires, business models, competition, and economies. Yet digitalization also provides possibilities for streamlined systems, new digital goods and services, and new business models powered by data. But the established business practices are no longer sufficient, as in the energy industry [48]. This business is at the cutting edge of several disruptive changes. The elevated ratio of clean energy and decentralized power production results in higher volatility and wider distribution (distributed energy consists of a range of smaller-scale and modular devices along with storage devices designed to supply energy to customers in the vicinity). The cycle with electricity as a product is done. Consumers deserve to be involved and want producers to take care of their particular preferences and beliefs.

ICT and IoT will play an important role in an unleashed energy world, and they will be vital elements for an efficient and advanced energy management system. Cybersecurity is becoming more vital with any further digitalization. So significant attention should be paid to any private communications system for utility services. The technological task is to put together the diverse specifications of the electrical and electronic environment, but this also requires a shift in culture and in people's digital capability. Utilities need to redesign themselves by means of digitalization tools if they are to continue in the central position of the sector.

10.3.12 Internet of Things

Promising applications for blockchain systems can indeed be described in IoT devices and ICT development, such as those in smart homes [49]. Blockchain implementations may promote IoT ecosystems, whereas open-source, public, and distributed blockchain systems would guarantee IoT interoperability requests [50].

IoT offers a forum for communicating with the electronic portion's essential component: the electricity and transmission system. IoT binds the energy-containing physical "battery" to the cloud network and then can be used to communicate with the applications. IoT often enables the conversion of energy via the physical medium upon the user's order by the device program. Blockchain is being used to make the whole thing decentralized, which helps connect all consumers mostly in a system from which they can operate without knowing one another or having any intermediary obligations. Decentralization aims to create a framework that needs no intermediary.

10.3.13 Transparent Process

Blockchain should be investigated for different implementations in the energy industry to address problems relating to encryption and transparency and to improve process output via the creation of a democratic definition of power, thus establishing a win-win scenario for all involved parties.

Blockchain would also facilitate consumers to participate throughout the entire electricity sector. They would depend on technology to make transactions faster, easier, and more affordable than a traditional centralized energy economy. Blockchain has also been assessed in electric vehicle (EV) charging installations, in which all battery packs for EVs are accessible to drivers through the development of an infrastructure of EVs and charging centers and the establishment of a simple payment and effective mediation mechanism for all parties concerned. Several oil firms are already pursuing blockchain solutions in elements of supply systems to enhance efficiency and to grow assets' output vs. expenditure. Contract finance is also studied where blockchain is implemented to boost payment efficiency and volatility [51].

Rapid fast-paced developments in emerging technology and the rising market engagement in energy conservation provide major prospects for shifting to a low-emission electricity system. Distributed ledger technologies, particularly blockchain, have started to gain considerable prominence in rendering the energy market healthier, more open, and more productive. As stated in the last section, blockchain has been investigated for many applications in the energy field.

10.3.14 Imbalance Settlement

In the traditional system, the settlement of payments related to imbalances is complex. However, in blockchain, the settlement of such cases is easy as it provides access to a distributed ledger. Also, the overall cost involved in the sharing of energy is realized on a real-time basis.

It is always helpful to initially identify the relation between load and energy to recognize the existence of load imbalance in the system. If E_a denotes the load on a microgrid E_l and energy available in the system, the net power E_{net} can be expressed mathematically as given in Eqn. (10.1):

$$E_{net} = \sum_{l=1}^{n} E_l - \sum_{a=1}^{m} E_a \qquad (10.1)$$

E_l is the summation of n number of loads present in the system at any point of time, and E_a is the summation of m number of energy generation sources. Nonzero value for E_{net} indicates an imbalance in the system, whereas the zero value indicates that the system is balanced. In the case of an imbalanced system, the value of net power will suggest the settlement procedure for the imbalance. This is depicted in the expression:

$$E_{net} > 0, \; i.e. \sum_{l=1}^{n} E_l - \sum_{a=1}^{m} E_a = positive, \; indicates \; that \; power \; is \; deficit \quad (10.2)$$

$$E_{net} < 0, \; i.e. \sum_{l=1}^{n} E_l - \sum_{a=1}^{m} E_a = negative, \; indicates \; that \; power \; is \; surplus \quad (10.3)$$

From Eqn. (10.2), it is clear that, for deficit power, additional power is required from other sources to settle the imbalance, whereas surplus power, as shown in Eqn. (10.3), can be settled by selling power or transferring it to any other system.

Settling fluctuations in the energy market are one feature of blockchains that has gained considerable coverage. Long negotiation procedures, quantity actualizations, and confirmations are the main reasons for observed delays. By keeping back office operations to a minimum, blockchains could reduce time and money delays. It is important to track and monitor the energy produced and used in free and clear ledgers to speed payments for services rendered. DLT implementations will include the convergence of metering systems with blockchains, which may entail high costs. Therefore, with close real-time authorizations, the business process itself will be more straightforward and effective. The network allowed by blockchain will encourage trade between various parties, boost audibility and transaction credibility, reduce the likelihood of fraudulent activity (through providing protected data storage), and require interoperability by standardizing data formats throughout various organizations [52].

In addition, the usage of blockchain-enabled smart contracts in the form of an imbalance settlement would allow accurate monitoring of supplier, user, and associated real-time payments. Nonetheless, though many services and businesses have started to consider the usage of blockchain for the resolution of imbalances, the problems of latency and poor throughput (i.e., transactions completed every second) remain obstacles that need to be tackled. Another problem is that ex post balance transfers function mostly like an already produced or spent offshore financial device and do not promote improvements in real-time behavior (for example, consumers will not provide a real-time message to buy less, i.e., adapt their requirement to availability, one of the main goals of the "smart grid" dream).

Deploying blockchains for various applications will increase the effectiveness and efficiency in this domain, but many limitations need to be addressed. Also, with technology moving at an alarmingly fast pace and 6G around the corner, there is a need to understand the challenges of using 6G networks and the potential benefits. Table 10.2 summarizes the benefits and limitations of using blockchain in the energy domain.

Blockchain for 6G is expected to address a large number of operational issues of 6G networks. As explained in Table 10.2, the benefits and limitations can be further categorized based on their utilization levels. The categorization is depicted in Table 10.3 and can be visualized in Figure 10.3.

10.4 CHALLENGES FOR BLOCKCHAIN IMPLEMENTATION IN THE ENERGY SECTOR

There are many challenges for blockchain implementation in the energy sector. Some of the important challenges can be categorized as development cost,

Table 10.2 Blockchain Applications in the Energy Sector

S. No.	Applications of blockchain	Benefits	Limitations
1	Energy bills	Billing is automatic and faster; the prepaid system can be easily adopted; a token system like Wattcoin can be used.	Regulatory policies do not exist and need major review.
2	Microgrids	Offers decentralized power to clients, better utilization of local power enhanced load factor.	Continuous supply and reliability of power can be assured within certain limits.
3	Sales and marketing	Improved monitoring of transactions, easy access to various markets and their scope identification; client-to-client sale is easy.	Pattern identification for the consumption of energy for different consumers.
4	Trading	Smooth conducting of interior and exterior activities; reconciliation is easier; reporting is easy.	Different frameworks are required for various locations and systems.
5	Automation	Improved and fast access to data related to energy, its tariff, and other transaction details.	Requires additional hardware and software for upgradation.
6	Smart grid	Sharing and storage of information are easy, fast, and mistake-proof; removes third-party intervention.	The implementation of advanced computational capabilities could create complexities.
7	Grid management energy security	Resolves the problem of expensive networking, improves the reliability of transactions, and enhances data security and privacy.	Separate and specific blockchain aggregators are required for each layer of the smart grid.
8	Utilization of renewable energy sources	It allows for direct interactions between producers and consumers and reliable record renewable generation and transaction, thus making the system trustworthy and attractive.	Lower reliability as renewable energy is dependent on weather.
9	Resource sharing	Effective resource sharing solutions among various consumers, as well as sharing among multiple users.	Generation of detailed records for users and their access.
10	Competition	The easy realization for collection of energy costs, optimization of energy tariffs.	Smart contracts are required.

(Continued)

Table 10.2 (Continued) Blockchain Applications in the Energy Sector

S. No.	Applications of blockchain	Benefits	Limitations
11	Digitalization	Transactions are fast, more information is available, real-time and easy access, enhances network performance.	A huge investment is needed; new platform-based businesses are required. F
12	Internet of Things	Promotes IoT ecosystems, connects all consumers mostly in the system from which they can operate without knowing one another.	IoT binds the energy-containing physical "battery" to the cloud network.
13	Transparent process	Improved process output, a win-win scenario for all involved parties, increased payment efficiency and volatility.	Editing of information once stored is a difficult and cumbersome process.
14	Imbalance settlement	The faster settlement, accurate monitoring, real-time access, significant reduction in time, and money delays.	Limited promotion for improvements in real-time behavior of energy requirement that may lead to imbalance.

Table 10.3 Categorization of Blockchain Applications in the Energy Sector

S. No.	Applications of blockchain	Benefits	Limitations
1	Energy bills	High	Low
2	Microgrids	Medium	High
3	Sales and marketing	High	Low
4	Trading	High	Medium
5	Automation	High	Medium
6	Smartgrid	High	Low
7	Grid management energy security	Medium	Medium
8	Utilization of renewable energy sources	Medium	High
9	Resource sharing	Medium	Low
10	Competition	High	Medium
11	Digitalization	High	High
12	Internet of Things	High	Medium
13	Transparent process	High	Low
14	Imbalance settlement	High	Medium

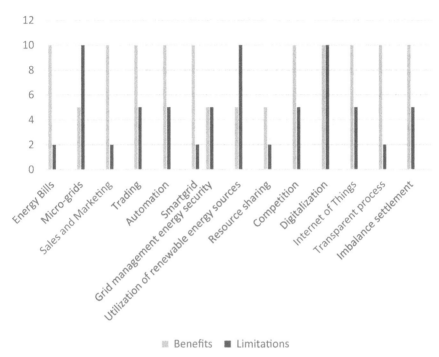

Figure 10.3 Categorization of blockchain application in the energy sector.

migration toward new technology, coordination with regulatory bodies, and lack of flexibility. The details of all these terms can be found in the following subsections.

10.4.1 Development Cost

Presently a major challenge for blockchain is its high development cost [53]. The energy data exchange cost for blockchain is low, but its validation and security costs are very high due to the required hardware and energy costs. Another factor that increases the initial cost of blockchain is data storage cost in dynamic ledgers. The cost of smart metering and the latest computerized setup is another expenditure for the development of blockchain. It might also need new infrastructures like software, customized equipment, enhanced security, and integrity of data-related expenditures. The integration of the metering system with the existing grid setup and distributed ledgers will place a further load on the developmental expenditure.

Some of the expenditures can be circumvented in blockchain, for example, the cost of intermediaries, using existing databases for storing energy records. Although blockchain has a huge potential to change the energy sector, due to high development expenditure and other challenges, it has a long way to travel.

10.4.2 Migration towards New Technology

Apart from cost, other hurdles for blockchain are adaptability and the face-off with the latest technologies. Various established technologies in this segment are already established and performing well. For example, in communication technology, telemetry is a good technique and cheap compared to blockchain. Peer-to-peer trading of energy is still undergoing advancement, and several technologies are emerging. Blockchain has to face all such techniques. In addition to competing with the technologies, blockchain's challenge is to balance, control, and coordinate with the technologies used in the grid [54].

10.4.3 Coordination with Regulatory Bodies

A key barrier for blockchain in the energy sector is coordination with the governing and regulatory bodies. In the energy market, policies framed by local regulatory bodies support the customers in providing energy at a cheaper cost and promoting renewable or low-carbon emission technologies. Blockchain can join such policies only after collaboration with local governing bodies, and such collaboration will need an amendment or new types of frameworks or contracts with the governing bodies. New frameworks may also be required as the existing frameworks may not include peer-to-peer trading and energy sharing between two independent consumers. Another big task is to settle the tariff of energy that is strongly regulated in the present scenario, even though a flexible tariff is needed for a blockchain system. Thus coordination and settlement of tariffs are additional requirements that need to be addressed in frameworks with regulatory bodies.

Presently the rules, like the privacy of consumer data and the like, are taken care of by the regulatory authorities. Blockchain systems will also need to recognize their users and simultaneously maintain their confidentiality. Its ledger has to maintain confidential data like identity as well the tariff agreed between suppliers and consumers. Additionally, the blockchain system will be required to follow legal and other obligations of the local authorities [55].

10.4.4 Lack of Flexibility

One more important element that blockchain has to take care of is the flexibility and standardization of its systems. The measures of the blockchain system are to be designed so that these are flexible enough for interoperability among various techniques. In the blockchain, the editing or updating of information at decentralized ledgers is very complex, which reduces its flexibility. Moreover, updates in ruling protocols require the consent of the system nodes. In a larger system like energy, such complexities may lead to a lack of interest among consumers and may hamper the reputation of the blockchain system [56].

10.5 CONCLUSION AND FUTURE DIRECTION

The chapter provides an exploratory analysis of blockchain uses in various energy applications and identifies the challenges encountered during blockchain implementation. Although the blockchain concept is now more widely identified with cryptocurrencies like bitcoin, its usefulness has been demonstrated in various ways and fields, through both academic research and real-life testing. As detailed in the chapter, it also provides a solution in the energy market. Blockchain might demonstrate a successful solution to several big problems, such as a push for decentralization/democratization, the need for more efficient setups, and increased resilience.

Blockchain may explicitly support processes, economies, and users of the energy sector. Though blockchain energy provides many benefits due to its improved protection, openness, and versatility, there are various drawbacks and threats both at the device level pertaining specifically to established parties in the energy market and at the application level itself. Pretty much across the board, blockchain technology is an exciting idea for the future of global, national, and foreign energy systems and also one where performance is heavily contingent on significant sociotechnical developments and perturbations of growth in the energy field. To impact these trajectories, either scholarly or professional proponents of blockchain technology need to integrate more science and advancement with realistic deployment and implementations in real life. In the future, continuous research is a must for the fast-growing blockchain technology. Existing individual use-cases can become a strong foundation for further research addressing the different applications and challenges occurring during the cases.

REFERENCES

1. Zhou, K., Yang, S., Shao, Z.: Energy internet: The business perspective (2016). https://doi.org/10.1016/j.apenergy.2016.06.052.
2. Wang, N., Zhou, X., Lu, X., Guan, Z., Wu, L., Du, X., Guizani, M.: When energy trading meets blockchain in electrical power system: The state of the art. *Appl. Sci.* 9, 1561 (2019). https://doi.org/10.3390/app9081561.
3. Gielen, D., Boshell, F., Saygin, D., Bazilian, M.D., Wagner, N., Gorini, R.: The role of renewable energy in the global energy transformation. *Energy Strateg. Rev.* 24, 38–50 (2019). https://doi.org/10.1016/j.esr.2019.01.006.
4. Andoni, M., Robu, V., Flynn, D., Abram, S., Geach, D., Jenkins, D., McCallum, P., Peacock, A.: Blockchain technology in the energy sector: A systematic review of challenges and opportunities (2019). https://doi.org/10.1016/j.rser.2018.10.014.

5. World Energy Outlook, www.iea.org/reports/world-energy-outlook-2017, last accessed 2020/09/13.

6. Tsao, Y.C., Thanh, V.V.: Toward blockchain-based renewable energy microgrid design considering default risk and demand uncertainty. *Renew. Energy.* 163, 870–881 (2021). https://doi.org/10.1016/j.renene.2020.09.016.

7. Sunyaev, A.: Distributed ledger technology. In: *Internet Computing.* Springer International Publishing, pp. 265–299 (2020). https://doi.org/10.1007/978-3-030-34957-8_9.

8. Dai, Y., Xu, D., Maharjan, S., Chen, Z., He, Q., Zhang, Y.: Blockchain and deep reinforcement learning empowered intelligent 5G beyond. *IEEE Netw.* 33, 10–17 (2019). https://doi.org/10.1109/MNET.2019.1800376.

9. Nguyen, T., Tran, N., Loven, L., Partala, J., Kechadi, M., Pirttikangas, S.: Privacy-aware blockchain innovation for 6G : Challenges and opportunities (2020). https://doi.org/10.1109/6GSUMMIT49458.2020.9083832.

10. Tahir, M., Habaebi, M.H., Dabbagh, M., Ahad, A.: A review on application of blockchain in 5G and beyond networks : Taxonomy, field-trials, challenges and opportunities (2020). https://doi.org/10.1109/ACCESS.2020.3003020.

11. Nofer, M., Gomber, P., Hinz, O., Schiereck, D.: Blockchain. *Bus. Inf. Syst. Eng.* 59, 183–187 (2017). https://doi.org/10.1007/s12599-017-0467-3.

12. Gupta, Sourav Sen.: Blockchain. *The Foundation Behind Bitcoin IBM Online.* (2017). http://www.IBM.COM).

13. Alam, T.: Blockchain and its role in the internet of things (IoT). *Int. J. Sci. Res. Comput. Sci. Eng. Inf. Technol.* 5, 151–157 (2019). https://doi.org/10.32628/cseit195137.

14. Blockchain Meets Energy, https://fsr.eui.eu/wp-content/uploads/Blockchain_meets_Energy_-_ENG.pdf, last accessed 2020/08/11.

15. Kumari, A., Gupta, R., Tanwar, S., Kumar, N.: Blockchain and A.I. amalgamation for energy cloud management: Challenges, solutions, and future directions. *J. Parallel Distrib. Comput.* 143, 148–166 (2020). https://doi.org/10.1016/j.jpdc.2020.05.004.

16. Blockchain in Energy and Utilities—Indigo Advisory Group | Strategy, Technology and Innovation, www.indigoadvisorygroup.com/blockchain, last accessed 2020/10/01.

17. Aiman, S., Hassan, S., Habbal, A., Rosli, A., Shabli, A.H.M.: Smart electricity billing system using blockchain technology. *J. Telecommun. Electron. Comput. Eng.* 10 (2019).

18. Dang, S.L., Yang, J.F., Wei, F.S., Que, H.K.: Design of prepayment implementation and its data exchange security protection based on power metering automatic system. In: *CSAE 2012 — Proceedings.* 2012 IEEE International Conference on Computer Science and Automation Engineering, pp. 331–335 (2012). https://doi.org/10.1109/CSAE.2012.6272966.

19. Ma, Q., Duan, C., Ding, X., Qian, T., Duan, P.: A design of network prepayment meter reading system based on ESAM. In: *Proceedings of the 2012 7th IEEE Conference on Industrial Electronics and Applications.* ICIEA, pp. 686–690 (2012). https://doi.org/10.1109/ICIEA.2012.6360813.

20. Lasseter, R.H., Paigi, P.: Microgrid: A conceptual solution. In: *PESC Record— IEEE Annual Power Electronics Specialists Conference*, pp. 4285–4290 (2004). https://doi.org/10.1109/PESC.2004.1354758.

21. Brooklyn Microgrid | Community Powered Energy, www.brooklyn.energy/, last accessed 2020/09/13.

22. Hou, W., Guo, L., Industrial, Z.N.-I.T. Undefined: Local electricity storage for blockchain-based energy trading in industrial internet of things (2019), ieeexplore.ieee.org.

23. Burger, C., Kuhlmann, A., Richard, P., Weinmann, J.: Blockchain in the energy transition. A survey among decision-makers in the German energy industry. *DENA Ger. Energy Agency*. 44 (2016).

24. Kounelis, I., Steri, G., Giuliani, R., Geneiatakis, D., Neisse, R., Nai-Fovino, I.: Fostering consumers' energy market through smart contracts. In: *Energy and Sustainability in Small Developing Economies, ES2DE 2017 — Proceedings*. Institute of Electrical and Electronics Engineers Inc. (2017). https://doi.org/10.1109/ES2DE.2017.8015343.

25. Tanwar, S., Kaneriya, S., Kumar, N., Zeadally, S.: ElectroBlocks: A blockchain-based energy trading scheme for smart grid systems. *Int. J. Commun. Syst.* 33, 1–15 (2020). https://doi.org/10.1002/dac.4547.

26. Blockchain Applications in Energy Trading, https://www2.deloitte.com/content/dam/Deloitte/global/Documents/Energy-and-Resources/gx-blockchain-applications-in-energy-trading.pdf, last accessed 2020/09/13.

27. Overview of Blockchain for Energy and Commodity Trading, www.ey.com/Publication/vwLUAssets/ey-overview-of-blockchain-for-energy-and-commodity-trading/%24FILE/ey-overview-of-blockchain-for-energy-and-commodity-trading.pdf, last accessed 2020/09/13.

28. Kumari, A., Shukla, A., Gupta, R., Tanwar, S., Tyagi, S., Kumar, N.: ET-DeaL: A P2P smart contract-based secure energy trading scheme for smart grid systems. IEEE INFOCOM 2020 — IEEE Conf. Comput. Commun. Work. INFOCOM WKSHPS 2020.1051–1056 (2020). https://doi.org/10.1109/INFOCOM WKSHPS50562.2020.9162989.

29. Kumari, A., Gupta, R., Tanwar, S., Tyagi, S., Kumar, N.: When blockchain meets smart grid: Secure energy trading in demand response management. *IEEE Netw.* 34, 299–305 (2020). https://doi.org/10.1109/MNET.001.1900660.

30. Tschorsch, F., Scheuermann, B.: Bitcoin and beyond: A technical survey on decentralized digital currencies. *IEEE Commun. Surv. Tutorials.* 18, 2084–2123 (2016). https://doi.org/10.1109/COMST.2016.2535718.

31. Narayanan, A.: *TechnoloBitcoin and Cryptocurrency Technologies: A Comprehensive Introduction.* Princeton University Press (2016).

32. Vukolić, M.: *Rethinking Permissioned Blockchains.* ACM Work (2017).

33. Alharby, M., van Moorsel, A.: Blockchain-based smart contracts : A systematic mapping study. arxiv.org. 125–140 (2017). https://doi.org/10.5121/csit.2017.71011.

34. Hassan, N.U., Member, S., Yuen, C., Niyato, D.: *Blockchain Technologies for Smart Energy Systems: Fundamentals.* Challenges and Solutions (2019).

35. Tanaka, K., Nagakubo, K., Abe, R.: Blockchain-based electricity trading with Digitalgrid router. In: *2017 IEEE International Conference on Consumer Electronics—Taiwan, ICCE-TW 2017*. Institute of Electrical and Electronics Engineers Inc. (2017), pp. 201–202. https://doi.org/10.1109/ICCE-China.2017. 7991065.

36. Sabounchi, M., Wei, J.: Towards resilient networked microgrids: Blockchain-enabled peer-to-peer electricity trading mechanism. In: *2017 IEEE Conference on Energy Internet and Energy System Integration, EI2 2017 — Proceedings*. Institute of Electrical and Electronics Engineers Inc. (2017), pp. 1–5. https://doi. org/10.1109/EI2.2017.8245449.

37. Mannaro, K., Pinna, A., Marchesi, M.: Crypto-trading: Blockchain-oriented energy market. In: *2017 AEIT International Annual Conference: Infrastructures for Energy and ICT: Opportunities for Fostering Innovation, AEIT 2017*. Institute of Electrical and Electronics Engineers Inc. (2017), pp. 1–5. https://doi. org/10.23919/AEIT.2017.8240547.

38. Li, Z., Kang, J., Yu, R., Ye, D., Deng, Q., Zhang, Y.: Consortium blockchain for secure energy trading in industrial internet of things. *IEEE Trans. Ind. Informatics.* 14, 3690–3700 (2018). https://doi.org/10.1109/TII.2017.2786307.

39. Mylrea, M., Gourisetti, S.N.G.: Blockchain: A path to grid modernization and cyber resiliency. In: *2017 North American Power Symposium, NAPS 2017*. Institute of Electrical and Electronics Engineers Inc. (2017). https://doi.org/10.1109/ NAPS.2017.8107313.

40. Su, W., Huang, A.Q.: *The Energy Internet: An Open Energy Platform to Transform Legacy Power Systems into Open Innovation and Global Economic Engines.* Elsevier (2018). https://doi.org/10.1016/C2016-0-04520-5.

41. Musleh, A.S., Yao, G., Muyeen, S.M.: Blockchain applications in smart grid-review and frameworks. *IEEE Access.* 7, 86746–86757 (2019). https://doi. org/10.1109/ACCESS.2019.2920682.

42. Brilliantova, V., Thurner, T.W.: Blockchain and the future of energy. *Technol. Soc.* 57, 38–45 (2019). https://doi.org/10.1016/j.techsoc.2018.11.001.

43. Sharma, P.K., Kumar, N., Park, J.H.: Blockchain technology toward green IoT: Opportunities and challenges. *IEEE Netw.* 34, 263–269 (2020). https://doi. org/10.1109/MNET.001.1900526.

44. Ashley, M.J., Johnson, M.S.: Establishing a secure, transparent, and autonomous blockchain of custody for renewable energy credits and carbon credits. *IEEE Eng. Manag. Rev.* 46, 100–102 (2018). https://doi.org/10.1109/EMR.2018.2874967.

45. Xu, H., Klaine, P.V., Onireti, O., Cao, B., Imran, M., Zhang, L.: Blockchain-enabled resource management and sharing for 6G communications. *Digit. Commun. Networks.* 6, 261–269 (2020). https://doi.org/10.1016/j.dcan.2020.06.002.

46. Ketter, W., Collins, J., Reddy, P.: Power TAC: A competitive economic simulation of the smart grid. *Energy Econ.* 39, 262–270 (2013). https://doi.org/10.1016/j. eneco.2013.04.015.

47. Jaradat, M., Jarrah, M., Bousselham, A., Jararweh, Y., Al-Ayyoub, M.: The internet of energy: Smart sensor networks and big data management for smart grid. *Procedia Comput. Sci.* 592–597. Elsevier B.V. (2015). https://doi.org/10.1016/j. procs.2015.07.250.

48. Varela, I.: Energy is essential, but utilities? Digitalization: What does it mean for the energy sector? In: *Digital Marketplaces Unleashed*. Springer, pp. 829–838 (2017). https://doi.org/10.1007/978-3-662-49275-8_73.

49. Risteska Stojkoska, B.L., Trivodaliev, K. V.: A review of internet of things for smart home: Challenges and solutions (2017). https://doi.org/10.1016/j.jclepro.2016.10.006.

50. Dhanji, T., Modi, C.: *Overview of Blockchain for Energy and Commodity Trading* (2017).

51. Khatoon, A., Verma, P., Southernwood, J., Massey, B., Corcoran, P.: Blockchain in energy efficiency: Potential applications and benefits, mdpi.com. https://doi.org/10.3390/en12173317.

52. Kabalci, Y.: A survey on smart metering and smart grid communication. *Renew. Sustain. Energy Rev.* Elsevier (2016). https://doi.org/10.1016/j.rser.2015.12.114.

53. Zheng, Z., Xie, S., Dai, H.N., Chen, X., Wang, H.: Blockchain challenges and opportunities: A survey. *Int. J. Web Grid Serv.* 14, 352–375 (2018). https://doi.org/10.1504/IJWGS.2018.095647.

54. Miglani, A., Kumar, N., Chamola, V., Zeadally, S.: Blockchain for internet of energy management: Review, solutions, and challenges (2020). https://doi.org/10.1016/j.comcom.2020.01.014.

55. Bürer, M.J., de Lapparent, M., Pallotta, V., Capezzali, M., Carpita, M.: Use cases for blockchain in the energy industry opportunities of emerging business models and related risks. *Comput. Ind. Eng.* 137, 106002 (2019). https://doi.org/10.1016/j.cie.2019.106002.

56. Ahl, A., Goto, M., Yarime, M., Tanaka, K., Sagawa, D.: *Practical Challenges and Opportunities of Blockchain in the Energy Sector: Expert Perspectives in Germany* (2019).

Index